SpringerBriefs in Computer Science

SpringerBriefs present concise summaries of cutting-edge research and practical applications across a wide spectrum of fields. Featuring compact volumes of 50 to 125 pages, the series covers a range of content from professional to academic.

Typical topics might include:

- A timely report of state-of-the art analytical techniques
- A bridge between new research results, as published in journal articles, and a contextual literature review
- A snapshot of a hot or emerging topic
- An in-depth case study or clinical example
- A presentation of core concepts that students must understand in order to make independent contributions

Briefs allow authors to present their ideas and readers to absorb them with minimal time investment. Briefs will be published as part of Springer's eBook collection, with millions of users worldwide. In addition, Briefs will be available for individual print and electronic purchase. Briefs are characterized by fast, global electronic dissemination, standard publishing contracts, easy-to-use manuscript preparation and formatting guidelines, and expedited production schedules. We aim for publication 8–12 weeks after acceptance. Both solicited and unsolicited manuscripts are considered for publication in this series.

**Indexing: This series is indexed in Scopus, Ei-Compendex, and zbMATH **

Xiaoqiang Zhu • Yuan Liu • Chunpeng Wang

Intelligent Localization for Integrated Sensing and Communication

Machine Learning-Driven Approaches

Xiaoqiang Zhu
School of Cyberspace Science
and Technology
Beijing Jiaotong University
Beijing, China

Yuan Liu
College of Intelligent and Computing
Tianjin University
Tianjin, China

Chunpeng Wang
School of Cyber Security, Key Laboratory
of Computing Power Network
and Information Security
Qilu University of Technology
Jinan, China

ISSN 2191-5768 ISSN 2191-5776 (electronic)
SpringerBriefs in Computer Science
ISBN 978-981-96-9384-9 ISBN 978-981-96-9385-6 (eBook)
https://doi.org/10.1007/978-981-96-9385-6

This work was supported by China Postdoctoral Science Foundation GZC20230224 2024M750166 Talent Fund of Beijing Jiaotong University 2023XKRC016 Open Project Fund of Key Laboratory of Computing Power Network and Information Security 2023ZD039 National Natural Science Foundation of China 62401037.

© The Editor(s) (if applicable) and The Author(s) 2026. This book is an open access publication.

Open Access This book is licensed under the terms of the Creative Commons Attribution 4.0 International License (http://creativecommons.org/licenses/by/4.0/), which permits use, sharing, adaptation, distribution and reproduction in any medium or format, as long as you give appropriate credit to the original author(s) and the source, provide a link to the Creative Commons license and indicate if changes were made.

The images or other third party material in this book are included in the book's Creative Commons license, unless indicated otherwise in a credit line to the material. If material is not included in the book's Creative Commons license and your intended use is not permitted by statutory regulation or exceeds the permitted use, you will need to obtain permission directly from the copyright holder.

The use of general descriptive names, registered names, trademarks, service marks, etc. in this publication does not imply, even in the absence of a specific statement, that such names are exempt from the relevant protective laws and regulations and therefore free for general use.

The publisher, the authors and the editors are safe to assume that the advice and information in this book are believed to be true and accurate at the date of publication. Neither the publisher nor the authors or the editors give a warranty, expressed or implied, with respect to the material contained herein or for any errors or omissions that may have been made. The publisher remains neutral with regard to jurisdictional claims in published maps and institutional affiliations.

This Springer imprint is published by the registered company Springer Nature Singapore Pte Ltd.
The registered company address is: 152 Beach Road, #21-01/04 Gateway East, Singapore 189721, Singapore

If disposing of this product, please recycle the paper.

Preface

Indoor localization is fundamental to a wide range of IoT applications and critical to Integrated Sensing and Communication (ISAC) systems. As ISAC evolves, achieving **more efficient** data collection, **more intelligent** system adaptability, and **more accurate** localization results have become essential for enhancing situational awareness, optimizing resource allocation, and improving overall service quality. While satellite-based systems such as the Global Positioning System (GPS) are extensively used for outdoor localization, they become unreliable or unavailable in specific scenarios like indoor spaces, where signal blockages from walls and other obstacles hinder performance. Developing advanced indoor localization technologies is therefore essential for fully leveraging the capabilities of ISAC in 6G networks and addressing the varied requirements of contemporary applications.

Current indoor localization technologies face significant challenges in these areas. Traditional methods for collecting data, especially for Channel State Information (CSI), often require substantial human intervention, which limits scalability and efficiency. Additionally, many existing systems cannot intelligently adapt to rapidly changing environments, such as user movements or new obstacles, leading to suboptimal performance. Achieving high accuracy remains another challenge, particularly in indoor spaces where multipath propagation and signal interference can distort results. These shortcomings highlight the need for new approaches that can address these issues in a more efficient, intelligent, and accurate manner.

This book explores the use of CSI and machine learning to tackle the three key challenges of indoor localization:

1. **More Efficient CSI Collection**: Reducing human intervention in the data collection process through automated methods while ensuring high-quality, reliable data.
2. **More Intelligent CSI Updates**: Developing adaptive mechanisms that allow for real-time updates of CSI values, ensuring system robustness and flexibility in dynamic environments.

3. **More Accurate Localization Applications**: Employing advanced machine learning algorithms to improve localization precision, even in complex indoor settings.

In this book, we aim to provide a comprehensive overview of recent advancements and theoretical insights into CSI-based indoor localization systems. Various machine learning-based approaches are discussed and validated through extensive real-world experimental studies, highlighting their robustness and adaptability in complex indoor scenarios.

We believe this book is an essential resource for researchers, engineers, and practitioners seeking to develop advanced, intelligent indoor localization systems. It is suitable for students new to the field, as well as professionals and researchers looking to deepen their understanding or implement cutting-edge localization solutions.

Beijing, China	Xiaoqiang Zhu
Tianjin, China	Yuan Liu
Jinan, China	Chunpeng Wang

Acknowledgements Writing this book has been a challenging yet rewarding journey, and it would not have been possible without the support and guidance of many individuals. First and foremost, I would like to express my heartfelt gratitude to Prof. Chunpeng Wang (Qilu University of Technology) and Dr. Yuan Liu (Tianjin University), whose mentorship and insightful feedback have been invaluable throughout the development of this book. Their expertise and encouragement inspired me to explore new perspectives and overcome obstacles during the writing process.

I am deeply grateful to my colleagues at Beijing Jiaotong University for their unwavering support and collaboration. Special thanks go to Dalin Zhang, Tao Zhang, Nan Wang, and Yingying Yao for their constructive discussions, technical insights, and valuable suggestions, which significantly enriched the content of this book. I would also like to thank my students and research group members, including Mingbo Zhang, Siyu Hu, Chengyi Ren, Haowen Zhang, Wang Yu, and Xuanqi He, whose curiosity and dedication fueled many of the ideas and experiments discussed in this work. Their enthusiasm and hard work have been a constant source of motivation for me.

Additionally, I am profoundly thankful to my family and friends for their patience, understanding, and unconditional support. Their belief in me and my work gave me the strength to complete this book. I extend my gratitude to the editorial team at Springer Nature for their professionalism and guidance in bringing this book to publication.

To all those who have supported me, directly or indirectly, I offer my sincere thanks. This book is as much a reflection of your contributions as it is of my efforts.

This work is supported in part by the National Natural Science Foundation of China (62401037); Talent Fund of Beijing Jiaotong University (2023XKRC016); China Postdoctoral Science Foundation (GZC20230224, 2024M750166); Open Project Fund of Key Laboratory of Computing Power Network and Information Security (2023ZD039). Their generous support is gratefully acknowledged.

Competing Interests The authors have no competing interests to declare that are relevant to the content of this manuscript.

Contents

1 What Is Intelligent Indoor Localization Technology? 1
2 Machine Learning for CSI-Based Localization 25
3 Efficient Offline Data Collection 47
4 Intelligent Offline Data Updating 81
5 Accurate Online Data Application 113
6 Conclusion 147

Chapter 1
What Is Intelligent Indoor Localization Technology?

Abstract Indoor localization addresses the limitations of GPS in enclosed environments, where signal obstructions and complex spatial configurations hinder traditional positioning systems. Chapter 1 introduces intelligent indoor localization, highlighting its unique challenges and opportunities. It explores the differences between indoor and outdoor positioning, focusing on solutions to improve accuracy, adaptability, and intelligence. The chapter also highlights advanced technologies like machine learning and outlines the book's focus on three key areas: **efficient** Channel State Information (CSI) collection, **intelligent** CSI updates, and **accurate** localization. These themes aim to reduce manual effort, enhance real-time adaptability, and improve accuracy for impactful applications.

The chapter also emphasizes the significance of CSI in wireless communications, demonstrating its ability to capture detailed signal properties beyond other localization signals. Comparing traditional techniques such as trilateration, triangulation, and fingerprinting, underscores CSI's advantages in addressing complex indoor scenarios. By providing a cohesive understanding of CSI's role, this chapter establishes a foundation for the development of robust, intelligent localization systems and serves as a roadmap for exploring the challenges and innovations presented throughout the book.

Keywords Indoor localization · Channel state information · Intelligent localization systems

1.1 Indoor Localization

1.1.1 How Is It Different from Outdoor Positioning?

Positioning technology is integral to the Internet of Things (IoT) and a key component of Integrated Sensing and Communication (ISAC) systems, supporting various applications (Zhu et al. 2020, 2025; Wang et al. 2022). While GPS has become the standard for outdoor positioning, its effectiveness is severely limited in indoor environments due to signal blockages from structural barriers such as walls and ceilings. This limitation has driven the development of indoor localization technologies specifically designed to address these constraints. Indoor and outdoor positioning distinctions lie primarily in their operating conditions and technical requirements (Liu et al. 2022; Cheng et al. 2022). Outdoor systems benefit from unobstructed signal propagation, which enables direct communication with satellites and supports large-scale coverage. Indoor environments, on the other hand, are characterized by challenges such as multipath propagation, signal attenuation, and interference caused by obstacles within enclosed spaces. These conditions necessitate innovative approaches that can reliably function under complex spatial configurations (Roy and Chowdhury 2021).

Moreover, the objectives of indoor localization systems differ significantly. Outdoor positioning focuses on broad accessibility and widespread applicability, whereas indoor localization prioritizes achieving high accuracy in confined areas. The dynamic and heterogeneous nature of indoor settings, with frequent layout changes and human movement, introduces further complexity to the task (Asaad and Maghdid 2022). To meet these challenges, intelligent indoor localization systems must fulfill three fundamental requirements:

- **Collection**: Efficiently gathering high-quality signal data that accurately reflects the characteristics of the indoor environment.
- **Update**: Dynamically adapting localization data to account for changes in the environment, such as layout modifications or variations in human activity.
- **Application**: Accurately utilizing the collected and updated data to deliver precise and reliable positioning outcomes tailored to real-world scenarios.

By addressing these requirements, intelligent indoor localization technologies are poised to overcome traditional limitations and advance the role of positioning in IoT and ISAC systems.

1.1.2 Overview of Intelligent Indoor Localization

Intelligent indoor localization offers an innovative solution to positioning challenges, achieving adaptive, high-precision results without requiring expensive external equipment. By leveraging technologies such as machine learning (ML) and intelligent algorithms, this approach directly addresses key issues in traditional positioning systems, including multipath effects, security vulnerabilities, and adaptability in dynamic environments (Guo et al. 2021). While existing localization methods–ranging from active and passive systems to device- or mobile-based and fingerprint-based techniques–have provided essential advancements, they often fall short in meeting the demands of complex scenarios. To tackle these limitations, this framework incorporates self-adaptive and self-learning capabilities, providing a robust foundation for advanced localization tasks.

1.1 Indoor Localization

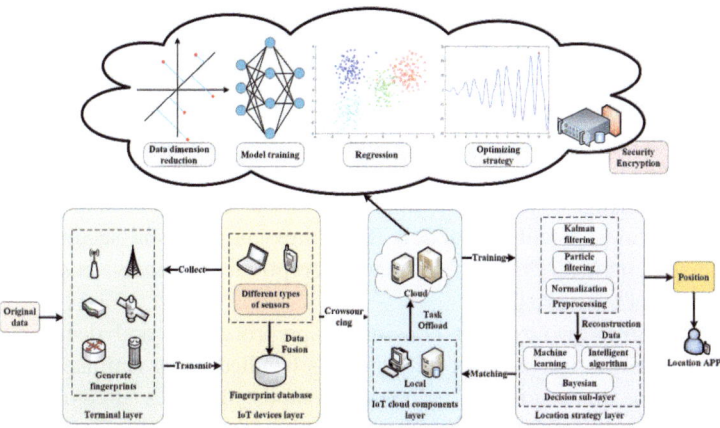

Fig. 1.1 Intelligent localization architecture

We propose the intelligent localization framework based on a four-layer architecture: the **terminal layer**, **IoT device layer**, **IoT cloud component layer**, and **location strategy layer**, complemented by cloud services and a user-centric localization application. Each layer is designed to seamlessly interact with the others to achieve efficient, secure, and scalable localization, as shown in Fig. 1.1 (Zhu et al. 2020).

1. *Terminal Layer.* It is responsible for sensing and collecting raw data from diverse sources such as satellites, routers, access points (APs), base stations, visible light systems, and various sensors. These signals form the basis for generating location fingerprints, which are captured by devices equipped with multiple sensing technologies, such as mobile phones and laptops. To unify heterogeneous signals, data fusion techniques are employed, establishing a consolidated fingerprint database that serves as the foundation for subsequent processing.
2. *IoT Device Layer.* It acts as the intermediary between the terminal layer and cloud services, facilitating the efficient transfer of collected data. Using crowdsourcing technology, the layer aggregates location fingerprints while optimizing data quality and reducing collection time. In addition, the IoT device layer ensures preprocessing capabilities, such as initial noise filtering, lightweight data compression, and data format standardization, before transferring the refined dataset to the IoT cloud component layer. These preprocessing steps alleviate the computational load on cloud resources and improve overall system responsiveness.
3. *IoT Cloud Component Layer.* This layer is the computational core of the framework, performing tasks such as dimension reduction, model training, regression problem-solving, and strategy optimization. Dimension reduction techniques, including Principal Component Analysis (PCA), mitigate the effects of noise and heterogeneity in large datasets, significantly reducing computational

complexity and training time. For model training, the framework adopts the Broad Learning System (BLS), an effective approach to addressing nonlinear localization challenges. BLS can be combined with other ML algorithms to minimize training errors and improve positioning accuracy. The layer also handles regression tasks through algorithms like k-Nearest Neighbors (KNN), which refine location predictions. Furthermore, optimization strategies address a range of challenges, including optimal dataset selection, signal source placement, and data dimensionality reduction. To address privacy and security concerns, this layer employs chaos theory and encryption algorithms, ensuring robust protection of sensitive data.

4. *Location Strategy Layer.* It consists of preprocessing and decision-making sublayers. Noise in the collected signals is handled using Kalman and particle filters, while dimensionality reduction methods like PCA streamline data for efficient processing. To address discrepancies caused by varying measurement accuracies across devices, normalization algorithms are employed. The sublayer also utilizes chaos theory for preprocessing nonlinear data, enhancing its quality and usability. The decision-making sublayer focuses on implementing advanced algorithms such as Bayesian probability to handle localization and trajectory tracking tasks. Since the position at any given moment is influenced by prior locations, trajectory tracking improves overall positioning accuracy. Results from this layer are integrated into a user-friendly application, providing actionable insights and services for end users.

This framework establishes a comprehensive system designed to meet the evolving needs of real-world applications. It highlights three primary requirements for intelligent localization: collection, ensuring accurate acquisition of heterogeneous signals; update, supporting adaptive and dynamic enhancements; and application, delivering actionable and precise localization results. This book believes that this framework will become the core of mainstream technologies.

1.1.3 Highlights of This Book

Intelligent indoor localization emphasizes adaptability and self-learning capabilities to tackle the increasing complexity of various scenarios. The objective is to develop systems that are more efficient, more intelligent, and more accurate, providing robust performance without relying on costly hardware or extensive manual intervention. This book focuses on three key aspects: Collection, Update, and Application, each addressing significant challenges in localization systems.

Collection involves acquiring high-quality, heterogeneous data from diverse signal sources such as Wi-Fi, Bluetooth, Radio Frequency Identification (RFID), visible light, and ultrasonic signals. Balancing the quantity and quality of data is a critical challenge. While coarse-grained data is easily accessible, it often lacks the precision needed for accurate localization, whereas fine-grained data collection frequently requires substantial manual effort or specialized hardware.

This book explores data fusion techniques to create a unified fingerprint database from multiple signal sources. Crowd-sensing methods are introduced to enhance scalability and reduce costs, but they also raise concerns about data reliability. Machine learning techniques, including clustering algorithms and anomaly detection models, are employed to identify and filter high-quality data, ensuring the fingerprint database is comprehensive and dependable.

Update ensures the system adapts dynamically to the ever-changing indoor environments, which are often subject to signal interference, user movement, and structural modifications. Static localization models struggle to maintain accuracy under such variability, making real-time adaptability essential for high-performance localization systems.

This book introduces adaptive techniques driven by reinforcement learning and transfer learning. Reinforcement learning enables the system to learn optimal strategies through continuous interaction with the environment, addressing changes such as signal fluctuations or newly introduced obstacles. Meanwhile, transfer learning allows pre-trained models to be efficiently fine-tuned for new settings, reducing the computational cost and training time. These methods work together to ensure that the system remains resilient and consistently delivers accurate localization in dynamic and unpredictable environments.

Application focuses on translating localization results into actionable insights that meet the demands of high accuracy and low latency. Scenarios such as emergency response, indoor navigation, and asset tracking require precise and timely results, yet achieving this often entails trade-offs between computational efficiency and accuracy.

This book proposes a multi-layered decision-making framework supported by machine learning. Preprocessing methods, such as PCA and noise reduction, optimize raw data for model processing. Decision-making algorithms, including KNN, BLS, and Bayesian probability models, are employed to refine predictions and enhance localization precision. Additionally, privacy-preserving mechanisms based on chaos theory and encryption algorithms are incorporated to safeguard data throughout the localization process. These innovations enable the framework to address the stringent requirements of intelligent localization applications.

1.2 Why Is Channel State Information?

1.2.1 Introduction to CSI in Wireless Communications

In wireless communication systems, accurately characterizing the radio channel is essential to maintain reliable signal transmission and reception. Among the many metrics available, CSI stands out as one of the most informative and critical

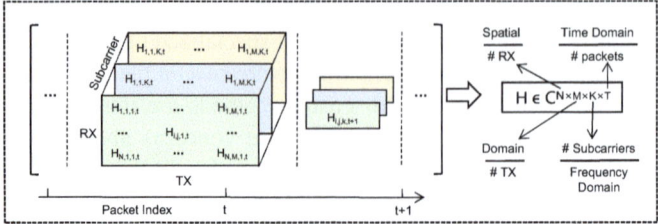

Fig. 1.2 The CSI architecture

indicators of channel properties. Unlike traditional metrics such as the Received Signal Strength Indicator (RSSI), which only measures signal strength, CSI provides a comprehensive view of the wireless channel by capturing fine-grained characteristics. These characteristics include the amplitude, phase, and time delay of individual subcarriers, which reflect the complex propagation and fading effects that occur in wireless environments, as shown in Fig. 1.2. Such detailed information allows researchers and engineers to analyze and optimize communication systems with a higher degree of precision.

The extraction of CSI provides access to detailed channel response information, enabling the study and understanding of precise signal characteristics. One of CSI's key strengths lies in its robustness against interference, particularly when coupled with orthogonal frequency-division multiplexing (OFDM). OFDM divides the communication channel into multiple orthogonal subcarriers, effectively reducing the adverse impact of multipath effects. This makes CSI an ideal tool for various applications, including fingerprint-based indoor localization systems, where high accuracy is required. By leveraging the unique properties of CSI, these systems can achieve significantly improved performance in positioning accuracy compared to traditional approaches.

The transmission process of a wireless signal, as represented by CSI, can be modeled mathematically as follows:

$$\mathbf{R} = CSI \cdot \mathbf{T} + \mathbf{G}, \tag{1.1}$$

where $\mathbf{T}$ is the transmitted signal vector, $\mathbf{G}$ represents Gaussian noise, and CSI is the channel matrix encapsulating the wireless channel's characteristics. The dimensions of the CSI matrix depend on the number of subcarriers, each subcarrier characterized by its frequency response. For the ith subcarrier, CSI can be expressed as a complex value:

$$CSI_i = |CSI_i| e^{j \angle CSI_i}, \tag{1.2}$$

where $|CSI_i|$ denotes the amplitude and $\angle CSI_i$ represents the phase. It serves as a robust tool for constructing fingerprint databases for indoor localization. Unlike metrics like RSSI, which are prone to variations due to environmental changes,

1.2 Why Is Channel State Information?

Fig. 1.3 Device schematic diagram

fingerprints derived from CSI demonstrate significant spatial variation and remain stable under dynamic conditions. This makes them particularly effective for high-precision localization tasks. Furthermore, the orthogonality of subcarriers in OFDM allows CSI-based systems to minimize interference, ensuring enhanced reliability and accuracy even in complex indoor environments.

The collection of CSI requires specialized hardware capable of measuring the amplitude and phase of individual subcarriers in real time. Such hardware enables the extraction of detailed channel characteristics, facilitating advanced research in both wireless communication and localization systems. Several key hardware options commonly utilized for CSI data collection are outlined below, as shown in Fig. 1.3:

- **Intel 5300 NIC:** This is one of the most widely used tools in academic and industrial research. Supporting the 802.11n Wi-Fi standard, it allows researchers to access per-packet CSI data through driver modifications. And it's particularly suited for controlled experiments and has seen extensive application in indoor localization systems due to its reliability and accessibility.
- **Atheros Wi-Fi Chipsets:** Though less flexible than the Intel 5300 NIC, these chipsets also offer CSI measurement capabilities. Often used in cost-sensitive or commercial systems, they require specialized drivers to extract CSI data. Despite their limitations, they provide a practical solution for applications where budget and simplicity are critical.
- **Software-Defined Radios (SDR):** SDRs, such as the Universal Software Radio Peripheral (USRP), stand out for their high programmability and support for diverse wireless protocols, including Wi-Fi, LTE, and 5G. These devices are ideal for experimental environments, offering unmatched flexibility compared to conventional Wi-Fi NICs. Their versatility makes them invaluable for investigating a wide range of wireless communication scenarios.
- **Specialized Localization Devices:** Platforms such as Ubiquiti Networks' AirMax and Qualcomm's Snapdragon series are designed to enhance CSI logging capabilities. These devices, optimized for large-scale deployments, provide real-time channel estimation and support high-precision localization in dynamic and challenging environments.

Despite the significant advantages offered by these devices, the primary challenge lies in ensuring the extraction of accurate CSI data. This process requires precise synchronization of signal transmission and reception, rigorous calibration to correct

Table 1.1 Comparison of hardware for CSI collection

Device	Features	Advantages	Limitations
Intel 5300 NIC	Supports 802.11n standard; driver modifications enable CSI extraction.	Widely used; ideal for controlled experiments; low-cost.	Limited to 802.11n standard; struggles with large-scale deployments.
Atheros Wi-Fi Chipsets	Provides CSI with specialized drivers; supports Wi-Fi standards.	Cost-effective; suitable for commercial systems.	Less flexible; limited research applications.
SDR	Collects CSI across protocols (Wi-Fi, LTE, 5G); highly programmable.	Versatile; ideal for experimental setups; supports multiple protocols.	Expensive; requires advanced programming skills.
Specialized Localization Devices	Platforms like AirMax and Snapdragon; optimized for real-time estimation.	Excellent for large-scale deployments; high accuracy.	Limited customization; high setup complexity.

measurement errors, and the application of noise-filtering techniques. Furthermore, in large-scale applications, managing the substantial volume of data generated by CSI measurements becomes crucial to maintaining system efficiency. A comparative overview of these hardware options is presented in Table 1.1.

CSI as the fine-grained data is particularly important for advanced technologies such as Multiple Input Multiple Output (MIMO), beamforming, and OFDM, all of which rely on precise channel information to optimize data transmission. CSI enables adaptive transmission strategies like adaptive modulation and coding, which adjust transmission parameters based on real-time channel conditions, enhancing system throughput and reliability. Furthermore, it facilitates interference mitigation and allows for efficient resource allocation in dense or complex environments. By capturing detailed channel characteristics, CSI empowers wireless systems to improve their efficiency, reliability, and capacity, making it a cornerstone of advanced wireless communication technologies, including 5G and beyond. It has emerged as a powerful tool for improving localization accuracy, especially in complex indoor environments where traditional localization techniques, such as GPS, are not effective. The importance of CSI in localization lies in its ability to offer unique "fingerprints" for each location within a wireless communication network, which can then be used to precisely estimate the position of a device or user.

For intelligent indoor localization, the CSI has several unique superiority as following (Chen et al. 2023):

- **High Precision Localization:** Traditional localization methods, such as RSSI and trilateration, provide basic estimates based on signal strength or distance. However, these methods are highly susceptible to multipath effects, interference,

1.2 Why Is Channel State Information?

and other environmental factors. In contrast, CSI, which captures both amplitude and phase information across multiple subcarriers, offers a much more granular and robust representation of the signal propagation environment. This enables higher precision in determining the position of a device, even in the presence of obstacles, signal reflections, and multi-path propagation.

- **Multipath Propagation and Reflection Handling:** One of the biggest challenges in indoor localization is the effect of multipath propagation, where signals bounce off walls, floors, and other objects. Traditional methods struggle to deal with these reflections, leading to poor localization accuracy. However, CSI can distinguish between the direct path and multipath components of the signal, allowing localization systems to separate the contributions from different paths. This enables systems to estimate position more reliably by exploiting the detailed channel characteristics captured by CSI.
- **Fingerprinting for Localization:** In indoor localization, CSI-based fingerprinting has proven to be a promising approach. The channel characteristics at different locations within a building or indoor space are unique, and these characteristics can be mapped into a database, creating a "fingerprint" for each location. When a device receives CSI data, it can compare the observed fingerprint with a pre-existing database to determine its location. This method has been shown to outperform RSSI-based localization in terms of accuracy and robustness.
- **Real-Time Positioning:** The high temporal resolution of CSI allows real-time tracking and continuous localization. In dynamic environments, such as mobile networks or environments with moving objects, CSI can be updated in real-time, providing accurate and continuous localization information, making it suitable for applications such as autonomous navigation and indoor robotics.
- **Multi-User Localization:** CSI's fine-grained nature enables multi-user localization in environments where multiple devices or users are active simultaneously. This is particularly useful in environments like offices, shopping malls, or large indoor facilities where numerous devices need to be localized at once. By using advanced algorithms that exploit CSI data from multiple users, systems can achieve accurate localization even in crowded environments.
- **Low-Cost and Low-Infrastructure Requirements:** One of the major advantages of CSI-based localization is that it can often be implemented using existing Wi-Fi infrastructure, reducing the need for additional hardware or expensive infrastructure modifications. By utilizing CSI from off-the-shelf Wi-Fi routers and access points, CSI-based localization systems offer a cost-effective alternative to traditional localization technologies that require specialized hardware or GPS receivers.

By providing detailed and reliable channel information, CSI enables the accurate and efficient determination of location in complex and dynamic environments. As wireless communication technology continues to evolve, CSI-based localization is poised to become a key enabler of many applications, including indoor navigation, smart homes, and location-based services (Fig. 1.4).

Fig. 1.4 Schematic diagram of CSI and other signals

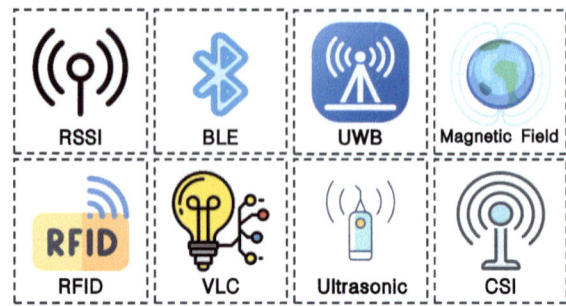

1.2.2 Comparisons Between CSI and Other Signals

RSSI

It's among the most widely adopted signals in wireless localization because of its straightforward acquisition process and minimal infrastructure requirements. Represented in dBm, RSSI measures the power level of a received signal, providing a quantitative indicator of signal strength at the receiver. This strength is influenced by factors such as transmission power, antenna gains, and propagation distance. Under ideal free-space conditions, the relationship between received signal strength and the distance between the transmitter and receiver is described by the Friis transmission equation:

$$P_r(d) = P_t \frac{G_t G_r}{(4\pi d/\lambda)^2}, \quad (1.3)$$

where $P_r(d)$ is the received power at distance d, P_t is the transmitted power, G_t and G_r are the antenna gains at the transmitter and receiver, respectively, and λ represents the signal wavelength. In real-world environments, especially indoors, the propagation of RSSI deviates significantly from this idealized model due to multipath effects, interference, and environmental obstructions. To better model signal attenuation under these conditions, the path loss model is often employed:

$$p(d) = p(d_0) - 10n \log_{10}\left(\frac{d}{d_0}\right) + G_\sigma, \quad (1.4)$$

where d_0 is the reference distance, $p(d_0)$ is the received power at d_0, n is the path loss exponent that reflects the environmental attenuation (commonly ranging from 2 to 5 in indoor scenarios), and G_σ represents random fluctuations modeled as a Gaussian variable with zero mean and a variance determined by environmental factors.

The primary appeal of RSSI-based methods lies in their simplicity, cost-effectiveness, and compatibility with existing wireless systems, making them a

1.2 Why Is Channel State Information?

practical choice for scenarios where infrastructure is limited or resource constraints are present. However, the accuracy of RSSI can be significantly impacted by environmental dynamics, such as signal occlusion by objects, human mobility, and varying atmospheric conditions like humidity or temperature. These factors often lead to inconsistencies between measured and actual signal strength, affecting localization performance. Despite these challenges, RSSI's flexibility has driven the development of advanced algorithms and correction techniques to improve its robustness. Adaptive methods, such as environment-aware calibration and machine learning models, are often applied to mitigate environmental impacts, enabling RSSI to deliver reliable performance in moderately controlled environments. These enhancements ensure that RSSI remains a viable option for many wireless localization applications, even in the face of its inherent limitations.

Bluetooth Low Energy (BLE)

It's a short-range wireless communication protocol designed for low-power and high-frequency applications. BLE operates in the 2.4 GHz ISM band, where it utilizes frequency-hopping spread spectrum (FHSS) across 40 channels to reduce interference and improve signal robustness (Shit et al. 2019). This feature makes BLE particularly suitable for applications such as indoor navigation, asset tracking, and proximity-based services. BLE localization primarily relies on beacon devices, which periodically broadcast unique identifiers alongside their signal strength. These signals are received by compatible devices, such as smartphones or specialized sensors, to estimate the distance between the transmitter and receiver. Unlike traditional RSSI-based localization systems, BLE's beacon infrastructure allows for a more structured deployment, enhancing localization scalability.

The distance between BLE devices is typically estimated as the same as the RSSI, and it also suffers from the multipath effects, device orientation, and environmental factors like walls, furniture, and human presence. These variabilities often lead to inaccuracies in distance estimation. One notable advantage of BLE over RSSI lies in its flexibility and energy efficiency. BLE beacons, often powered by small coin-cell batteries, can function for months or years without replacement, enabling widespread and low-maintenance deployments. Additionally, BLE supports a wide range of commercial applications, including personalized marketing in retail spaces and wayfinding in large indoor environments like airports or shopping malls. However, BLE signals suffer from challenges such as signal instability and susceptibility to interference from other 2.4 GHz devices, including Wi-Fi. Advanced algorithms, including machine learning-based regression models and signal smoothing techniques like particle filters, are often employed to mitigate these issues. Furthermore, hybrid localization systems that integrate BLE with technologies such as ultra-wideband (UWB) or Wi-Fi can achieve higher accuracy.

UWB

It's a wireless communication technology that operates over a broad frequency spectrum, typically exceeding 500 MHz, using extremely short-duration pulses for data transmission. Unlike narrowband systems, UWB's wide bandwidth allows it to achieve fine temporal resolution, enabling accurate time-of-flight (ToF) measurements. These ToF measurements are the foundation for its exceptional localization accuracy, which can reach cm-level precision. UWB signals are highly resilient to interference and can effectively mitigate the impact of multipath effects, making them well-suited for challenging indoor environments. Furthermore, UWB can penetrate obstacles like walls and furniture to maintain robust signal quality, enhancing its reliability in real-world scenarios (Yang et al. 2022).

While UWB offers superior accuracy compared to other technologies like BLE or RSSI-based systems, it comes with some trade-offs. The shorter transmission distance typically limited to a few tens of meters, can restrict its applicability in large-scale deployments. Additionally, UWB requires specialized hardware, such as anchors and tags, which may increase the cost of implementation compared to systems leveraging existing infrastructure like Wi-Fi or BLE. Nevertheless, UWB has found increasing adoption in applications requiring precise localization, such as asset tracking, autonomous robotics, and secure access systems. With the integration of UWB technology into consumer devices, including smartphones and wearables, its use is expanding beyond industrial applications to everyday scenarios such as contactless payments and proximity-based interactions.

Magnetic Field

Magnetic field-based localization harnesses the Earth's geomagnetic field to determine position, drawing inspiration from animals like homing pigeons that navigate using magnetic cues. This approach relies on the principle that magnetic field strength and direction vary across locations, forming unique spatial signatures that can be mapped for indoor positioning (Kusche et al. 2021). In indoor environments, factors such as metallic objects, electronic devices, and structural elements influence the magnetic field's distribution, creating both challenges and opportunities. Despite these complexities, magnetic fields serve as a dependable fingerprint for localization, requiring no additional infrastructure like Wi-Fi or Bluetooth. This makes the method particularly useful in scenarios where other signals are unreliable or unavailable.

The magnetic field produced by a point dipole source is modeled by

$$B(\mathbf{m}, \mathbf{r}) = \frac{\mu}{4\pi} \left[3 \frac{(\mathbf{m}, \mathbf{r})\mathbf{r}}{|\mathbf{r}|^5} - \frac{\mathbf{m}}{|\mathbf{r}|^3} \right], \tag{1.5}$$

where μ is the permeability of free space, $\mathbf{m}$ represents the magnetic moment of the dipole, and $\mathbf{r}$ is the vector from the source to the observation point, with

1.2 Why Is Channel State Information?

$|\mathbf{r}|$ denoting the distance between them. This passive, infrastructure-free approach offers significant advantages over systems requiring active transmitters, such as Wi-Fi or UWB. Its cost-effectiveness and independence from external signal sources make it ideal for environments where traditional technologies face limitations. However, magnetic field-based localization is sensitive to environmental changes, such as nearby magnetic interference or material composition, which can impact positioning accuracy. Integrating this technology with other systems, like Wi-Fi or UWB, helps overcome these limitations and enhances localization precision. As a result, magnetic field-based localization has shown promise in unique applications, including underground, underwater, and signal-shielded environments. When used as part of a multi-sensor fusion system, it significantly enhances robustness and reliability, addressing limitations inherent in standalone systems.

RFID

This technology operates by enabling communication between readers and tags using electromagnetic fields. RFID tags are categorized as passive or active. Passive tags lack an internal power source, relying on energy harvested from the reader's signal. This makes them lightweight, economical, and suitable for various applications. Active tags, on the other hand, are equipped with internal batteries, allowing for extended communication distances and more robust functionality but at a higher cost and larger size. Localization methods using RFID include signal property analysis, such as RSSI, phase shifts, or proximity detection. For example, proximity-based methods estimate a tag's location by determining which reader detects the strongest signal, offering a straightforward but lower-accuracy solution. Phase-based techniques, by contrast, leverage the phase difference of signals to provide more precise positioning, particularly in controlled environments (Tang et al. 2024).

RFID localization finds widespread use in sectors such as logistics, warehousing, and retail, owing to its ability to identify and monitor numerous items simultaneously. Its adaptability and seamless integration into existing infrastructures make it particularly effective for managing inventories and monitoring personnel movements. However, environmental factors like interference or metallic objects can affect RFID signal propagation, impacting its reliability. Additionally, the communication range of passive RFID systems is generally restricted to a few meters, limiting their suitability for large-scale applications. To address these challenges, RFID systems are increasingly combined with complementary technologies, such as inertial sensors or Wi-Fi, to enhance performance. These hybrid solutions maximize RFID's cost-effectiveness while improving accuracy and resilience, expanding its utility in diverse applications, including access management and dynamic asset location monitoring.

Visible Light Communication (VLC)

It's an emerging technology that utilizes light-emitting diodes (LEDs) as the communication medium, offering a promising alternative to traditional RF-based localization systems. VLC operates under the green communication concept, making it not only an efficient and sustainable solution but also an innovative one in the context of IoT applications. Unlike RF technologies, VLC positioning requires only light intensity data collected by photodetectors or photodiodes, eliminating the need for additional infrastructure such as transmitters and receivers typically required by Wi-Fi or Bluetooth systems. This approach allows for simple deployment, low cost, and minimal power consumption, making VLC an attractive solution for smart building applications, indoor navigation, and location-based services.

The received signal in VLC positioning systems can be represented as

$$r(t) = \alpha R_p x(t - \tau) + n(t), \tag{1.6}$$

where α is the attenuation factor of the LED, τ is the flight time of the light signal, R_p is the responsivity of the photodiode, and $n(t)$ is the shot noise in the system. VLC's key advantages lie in its ability to provide high-precision localization while avoiding the multipath effects that commonly affect RF-based technologies. Since the signal is based on light, it is immune to interference from physical obstructions, making it especially effective in controlled indoor environments. Furthermore, VLC can leverage existing lighting systems, such as the LED lighting found in many modern buildings, for seamless integration with minimal extra cost. However, the most significant limitation is its reliance on line-of-sight (LoS) between the light source and the receiver, which means that obstacles such as walls or furniture can block the signal. Additionally, VLC systems can be sensitive to fluctuations in lighting conditions, such as changes in ambient light or power fluctuations in the LED sources. However, the potential of VLC in specific applications–such as indoor localization in areas where RF signals are weak or unavailable, or environments where energy efficiency is a priority–remains high. By combining VLC with other positioning technologies, such as Wi-Fi or UWB, it is possible to mitigate some of these challenges and create a more robust and reliable indoor localization system.

Ultrasonic

It utilizes high-frequency sound waves (above 20 kHz) for distance measurement and localization. It works by transmitting an ultrasonic pulse from a transmitter to a receiver, which then measures the time it takes for the pulse to travel to the receiver and return. This ToF measurement, combined with the speed of sound in the medium, is used to calculate the distance between the transmitter and the receiver. Ultrasonic positioning is highly accurate, with sub-cm precision achievable in ideal conditions, and particularly useful for applications requiring high-precision

1.2 Why Is Channel State Information?

localization, such as robotics, asset tracking, and industrial automation (Cai et al. 2022).

The general model for ultrasonic distance measurement can be expressed as

$$d = \frac{c \cdot t}{2}, \qquad (1.7)$$

where d is the distance between the transmitter and the receiver, c is the speed of sound in the medium (typically air), and t is the time taken for the sound pulse to travel to the receiver and back. Ultrasonic technology offers several advantages, including high accuracy, low power consumption, and the ability to operate without complex infrastructure. It can be implemented with relatively inexpensive hardware, making it suitable for both large-scale and small-scale applications.

While ultrasonic localization offers high accuracy, it is subject to certain limitations. One significant drawback is that ultrasonic waves are subject to interference from environmental factors such as temperature, humidity, and obstacles in the path of the signal, which can affect the accuracy of the distance measurements. Furthermore, ultrasonic positioning systems typically require a LoS between the transmitter and receiver, meaning obstacles such as walls or furniture can obstruct the signal. These limitations can make ultrasonic systems less effective in dynamic, cluttered environments compared to technologies like RFID or Wi-Fi. Nonetheless, ultrasonic technology remains a viable solution for applications in controlled indoor environments, where its accuracy and ease of integration with other systems make it a valuable tool for high-precision localization. Additionally, combining ultrasonic with other positioning technologies, such as visual or magnetic field-based systems, can help mitigate these limitations and enhance overall performance.

Discussion

CSI stands out among localization technologies due to its ability to capture detailed amplitude and phase information across multiple subcarriers. This fine-grained representation allows CSI to effectively utilize multipath propagation, a common challenge for technologies like RSSI and BLE, to enhance localization accuracy. Unlike these signal strength-based approaches, which are prone to interference and environmental variability, CSI leverages rich channel characteristics to turn multipath effects into an advantage. Moreover, CSI achieves high precision without requiring specialized hardware by utilizing standard Wi-Fi access points and off-the-shelf network interface cards, making it both cost-effective and accessible.

Compared to other high-accuracy localization technologies such as UWB, VLC, and ultrasonic, CSI offers unique benefits. UWB provides centimeter-level precision but requires dedicated hardware and operates within a constrained distance. Similarly, VLC and ultrasonic are limited by line-of-sight requirements and sensitivity to environmental factors. In contrast, CSI is more robust in obstructed or dynamic environments thanks to its reliance on Wi-Fi infrastructure. Its capability to support multi-user localization and advanced techniques such as beamforming further distinguishes it from magnetic field and RFID-based systems, which often face hardware dependencies and environmental constraints. The comparison of these technologies with CSI is summarized in Table 1.2, these characteristics position CSI as a strong contender for advancing wireless localization technologies, combining precision, dependability, and adaptability.

1.2.3 Wireless Localization Comparisons

Trilateration

Trilateration is a distance-based localization technique widely employed in indoor positioning systems. Its principle includes determining the precise location of a target point by utilizing distance measurements between the target and multiple reference points (such as anchor nodes or base stations) with known coordinates (Albraheem and Alawad 2023). The effectiveness of this method hinges on high-precision distance measurement techniques, with Time of Arrival (ToA) and Time Difference of Arrival (TDoA) being the most commonly adopted approaches. ToA estimates distance by measuring the time a signal takes to propagate from a reference point to the target, whereas TDoA determines the target's position by analyzing the time differentials of signal arrivals at multiple reference points. Although both ToA and TDoA are time-based localization techniques that require at least three base stations with known locations, they impose stringent time synchronization requirements on devices, thereby increasing the operational complexity in real-world applications.

The core principle of trilateration is based on the geometric intersection of circles or spheres, as shown in Fig. 1.5. In a two-dimensional plane, suppose three reference points $A(x_1, y_1)$, $B(x_2, y_2)$, and $C(x_3, y_3)$ are given, along with a target point $P(x, y)$, whose distances to these three reference points are d_1, d_2, and d_3, respectively. According to geometric relationships, the position of the target point P should satisfy the follows:

$$\begin{cases} \sqrt{(x - x_1)^2 + (y - y_1)^2} = d_1 \\ \sqrt{(x - x_2)^2 + (y - y_2)^2} = d_2 \\ \sqrt{(x - x_3)^2 + (y - y_3)^2} = d_3 \end{cases} \quad . \tag{1.8}$$

Each equation represents a circle centered at a reference point with a radius equal to the measured distance. The target point P lies at the intersection of these circles. The system of equations can be solved using algebraic methods, such as the least squares method, or geometric approaches, the coordinates (x, y) of the target point P can be determined.

1.2 Why Is Channel State Information?

Table 1.2 Comparison of localization technologies

Technology	Key principle	Advantages	Limitations	Key applications
RSSI	Signal strength measurement	Simple, low cost, widely available, uses existing infrastructure.	Susceptible to multipath effects, interference, and environmental changes; limited accuracy.	Indoor navigation, asset tracking, basic proximity detection.
BLE	Bluetooth signal strength and proximity	Low energy consumption, widely available in consumer devices, easy deployment.	Limited precision due to environmental interference; relatively short communication range.	Smart home automation, location-based marketing, fitness and health tracking.
UWB	ToF and wide bandwidth	High precision (cm-level), robust against multipath effects, low interference.	Specialized hardware is needed, leading to higher costs and restricted operational distance.	Autonomous robotics, high-accuracy asset tracking, secure access systems.
Magnetic Field	Earth's geomagnetic variations	Infrastructure-free, cost-effective, works in signal-deprived environments like underwater or underground.	Sensitivity to environmental changes can reduce precision; resolution is inherently limited.	Underground navigation, tunnel localization, signal-limited environments.
RFID	Radio frequency tag identification	Cost-effective for asset tracking, effective for proximity-based applications.	Operates within a short range and depends on the availability of tags and readers, making it less effective for fine-grained positioning.	Supply chain management, inventory tracking, and security systems.

(continued)

Table 1.2 (continued)

Technology	Key principle	Advantages	Limitations	Key applications
VLC	Light intensity and photodiode reception	High precision, immunity to electromagnetic interference, uses existing LED lighting infrastructure.	Direct LoS is essential, with performance affected by ambient light conditions and limited coverage.	Indoor navigation in smart buildings, secure data communication, industrial automation.
Ultrasonic	Time-of-flight of sound waves	High accuracy, low cost, performs well in controlled environments.	LoS constraints and susceptibility to environmental factors like temperature and humidity limit its effectiveness.	Robotic navigation, warehouse automation, and close-proximity object tracking.
CSI	Amplitude and phase of wireless signals	High precision, robust to environmental interference, supports multipath exploitation, works with existing Wi-Fi.	Computational complexity and hardware constraints can affect real-time performance.	Smart homes, healthcare monitoring, autonomous vehicles, multi-user localization.

1.2 Why Is Channel State Information?

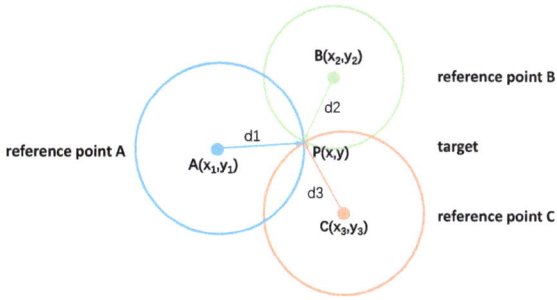

Fig. 1.5 Structural principle diagram of trilateration

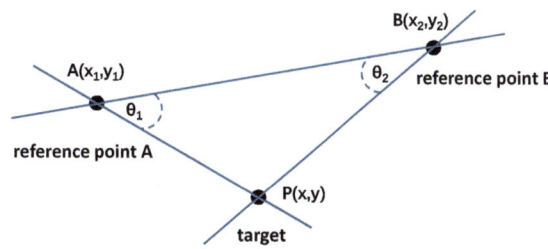

Fig. 1.6 Structural principle diagram of triangulation

Trilateration is a widely used positioning method based on geometric principles, well-suited for indoor environments with relatively stable signal conditions. However, its accuracy can be affected by multipath propagation, non-line-of-sight (NLOS) scenarios, and measurement errors. To improve performance under such conditions, practical systems often apply filtering algorithms or combine trilateration with other positioning techniques to achieve better accuracy and reliability.

Triangulation

Triangulation is another classic positioning method based on angle measurement, it determines the coordinates of a target by using angular information between known reference points and the target itself. Unlike trilateration, which depends on distance measurements, triangulation relies on directional angles, typically the Angle of Arrival (AoA), to estimate position. By measuring the angles at which signals from the target arrive at multiple reference points and applying geometric relationships, the target's location can be calculated. Because of its high accuracy and reliability, triangulation is widely used in radar systems, acoustic positioning, wireless communication networks, and indoor localization applications.

As shown in Fig. 1.6, consider a scenario where two reference points with known coordinates, $A(x_1, y_1)$ and $B(x_2, y_2)$, are able to measure the angles of arrival θ_1 and θ_2, respectively, from a target point $P(x, y)$. These angles are typically measured relative to a common axis (e.g., the x-axis), based on the direction of incoming signals. By applying basic trigonometric relationships, the coordinates of the target point can be expressed as:

$$\tan(\theta_1) = \frac{y - y_1}{x - x_1},$$

$$\tan(\theta_2) = \frac{y - y_2}{x - x_2}. \tag{1.9}$$

Solving this system of equations yields the explicit coordinates of the target point:

$$x = \frac{y_2 - y_1 + x_1 \tan(\theta_1) - x_2 \tan(\theta_2)}{\tan(\theta_1) - \tan(\theta_2)},$$

$$y = y_1 + (x - x_1) \tan(\theta_1). \tag{1.10}$$

This method enables accurate localization without requiring direct distance measurements, provided that the angular information is measured with sufficient precision. In practical scenarios, the effectiveness of triangulation is influenced by both the resolution of the angle estimation technique and the geometric configuration of the reference points. When the reference points are arranged in an unfavorable manner, for instance when they are nearly colinear, the resulting angular ambiguity and amplification of measurement errors can significantly reduce localization accuracy. Therefore, achieving reliable performance in real-world applications necessitates suitable system design and the use of advanced angle estimation methods, such as antenna arrays or directional sensing devices.

Fingerprint-Based Localization

It represents a different paradigm of indoor localization, and leverages the spatial variation characteristics of wireless signals to establish a mapping between signal features and spatial locations, enabling precise position estimation (Zhang et al. 2023). Unlike trilateration, which relies on distance measurements, or triangulation, which depends on angle measurements, fingerprinting does not require prior knowledge of the exact positions of base stations, nor does it rely on signal propagation time or angle measurements. This reduces the system's dependence on hardware synchronization and measurement precision, thereby enhancing its deployment flexibility and applicability. Because of its ease of implementation, low measurement cost, and strong adaptability to complex environments, fingerprinting has emerged as one of the mainstream techniques for indoor positioning.

The implementation process, as shown in Fig. 1.7, consists of two main phases: the offline phase and the online phase. In the offline phase, the target localization area is first divided into a uniform grid to define sampling points. Wireless fingerprint data, such as RSSI, are collected from multiple APs at these sampling locations. The multidimensional fingerprint data from each sampling point are then

1.2 Why Is Channel State Information?

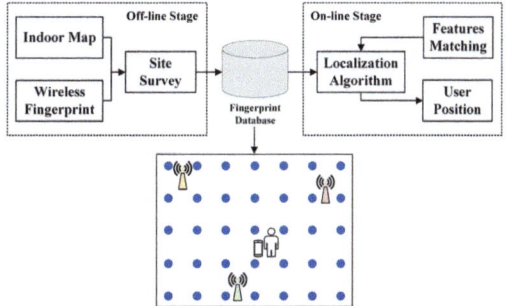

Fig. 1.7 Framework of WiFi fingerprint-based localization system

stored to construct an offline fingerprint database, a process commonly referred to as a site survey. In the online phase, when a user is at an unknown location, their device transmits real-time fingerprint data to the positioning algorithm. The system then searches the fingerprint database and identifies the most similar fingerprint records to estimate the user's position, thereby achieving localization. Common positioning algorithms include the KNN algorithm and machine learning-based classification or regression methods, all of which optimize the matching process to improve localization accuracy and robustness.

The advantage of fingerprinting-based localization lies in its strong adaptability to complex environments, effectively mitigating issues such as multipath effects and NLoS propagation. However, its offline phase requires extensive data collection, and environmental changes (e.g., furniture movement or variations in signal sources) may render the fingerprint database obsolete, necessitating recalibration. Therefore, in practical applications, this technology is often combined with ML algorithms or incremental update techniques to enhance the system's robustness and real-time performance (Ruan et al. 2022).

To summarize, the above sections analyze these three techniques comprehensively, with a comparisons provided in Table 1.3. Trilateration determines the position of a target point by measuring its distances from multiple reference points. This technique is widely applied in wireless sensor networks and other localization scenarios; however, it relies on favorable signal propagation conditions to ensure accuracy. In contrast, triangulation estimates position based on angle measurements and trigonometric computations. While effective in complex environments, it often requires more sophisticated and costly equipment. Fingerprinting, by leveraging multipath effects and environmental characteristics, establishes a fingerprint database and matches real-time signal features for localization. This approach mitigates the inherent errors of traditional ranging-based methods and offers superior localization accuracy in challenging environments.

Table 1.3 Wireless localization comparisons

Method	Principle	Advantages	Limitations	Applications
Trilateration	Calculating position based on distances from multiple reference points using ToA/TDoA	Simple theoretical model; does not rely on environmental characteristics	Highly sensitive to distance measurement errors; requires a well-calibrated signal propagation model	Indoor/outdoor environments with reliable distance measurements (e.g., UWB, LoRa)
Triangulation	Estimating position using angle measurements from multiple known reference points	Can achieve high accuracy in specific scenarios; widely used in optical and ultrasound-based systems	Requires precise angle measurements; errors accumulate over long distances	Applications requiring precise angular positioning, such as camera-based and LiDAR localization
Fingerprinting	Collecting and matching signal characteristics (e.g., RSSI) in a pre-built database	High adaptability to complex environments; robust against multipath effects and NLoS propagation	Requires extensive offline data collection; environmental changes may lead to fingerprint database degradation	Large indoor spaces such as malls, airports, and museums

References

Albraheem L, Alawad S (2023) A hybrid indoor positioning system based on visible light communication and bluetooth RSS trilateration. Sensors 23(16):7199

Asaad SM, Maghdid HS (2022) A comprehensive review of indoor/outdoor localization solutions in IoT era: research challenges and future perspectives. Comput Netw 212:109041

Cai C, Zheng R, Luo J (2022) Ubiquitous acoustic sensing on commodity IoT devices: a survey. IEEE Commun Surv Tuts 24(1):432–454

Chen C, Zhou G, Lin Y (2023) Cross-domain wifi sensing with channel state information: a survey. ACM Comput Surv 55(11):1–37

Cheng X, Huang Z, Bai L (2022) Channel nonstationarity and consistency for beyond 5G and 6G: a survey. IEEE Commun Surv Tuts 24(3):1634–1669

Guo H, Li J, Liu J, Tian N, Kato N (2021) A survey on space-air-ground-sea integrated network security in 6G. IEEE Commun Surv Tuts 24(1):53–87

Kusche R, Schmidt SO, Hellbrück H (2021) Indoor positioning via artificial magnetic fields. IEEE Trans Instrum Measur 70:1–9

Liu A, Huang Z, Li M, Wan Y, Li W, Han TX, Liu C, Du R, Tan DKP, Lu J, et al. (2022) A survey on fundamental limits of integrated sensing and communication. IEEE Commun Surv Tuts 24(2):994–1034

Roy P, Chowdhury C (2021) A survey of machine learning techniques for indoor localization and navigation systems. J Intell Robot Syst 101(3):63

References

Ruan Y, Chen L, Zhou X, Guo G, Chen R (2022) Hi-Loc: hybrid indoor localization via enhanced 5G NR CSI. IEEE Trans Instrum Measur 71:1–15

Shit RC, Sharma S, Puthal D, James P, Pradhan B, Van Moorsel A, Zomaya AY, Ranjan R (2019) Ubiquitous localization (UbiLoc): a survey and taxonomy on device free localization for smart world. IEEE Commun Surv Tuts 21(4):3532–3564

Tang P, Yin Y, Tong Y, Liu S, Li L, Jiang T, Wang Q, Chen M (2024) Channel characterization and modeling for VLC-IoE applications in 6G: a survey. IEEE Internet Things J 11:34872–34895

Wang J, Varshney N, Gentile C, Blandino S, Chuang J, Golmie N (2022) Integrated sensing and communication: enabling techniques, applications, tools and data sets, standardization, and future directions. IEEE Internet Things J 9(23):23416–23440

Yang B, Li J, Shao Z, Zhang H (2022) Robust UWB indoor localization for NLOS scenes via learning spatial-temporal features. IEEE Sens J 22(8):7990–8000

Zhang B, Sifaou H, Li GY (2023) CSI-fingerprinting indoor localization via attention-augmented residual convolutional neural network. IEEE Trans Wireless Commun 22(8):5583–5597

Zhu X, Qu W, Qiu T, Zhao L, Atiquzzaman M, Wu DO (2020) Indoor intelligent fingerprint-based localization: principles, approaches and challenges. IEEE Commun Surv Tuts 22(4):2634–2657

Zhu X, Liu J, Lu L, Zhang T, Qiu T, Wang C, Liu Y (2025) Enabling intelligent connectivity: a survey of secure ISAC in 6G networks. IEEE Commun Surv Tuts 27:748–781

Open Access This chapter is licensed under the terms of the Creative Commons Attribution 4.0 International License (http://creativecommons.org/licenses/by/4.0/), which permits use, sharing, adaptation, distribution and reproduction in any medium or format, as long as you give appropriate credit to the original author(s) and the source, provide a link to the Creative Commons license and indicate if changes were made.

The images or other third party material in this chapter are included in the chapter's Creative Commons license, unless indicated otherwise in a credit line to the material. If material is not included in the chapter's Creative Commons license and your intended use is not permitted by statutory regulation or exceeds the permitted use, you will need to obtain permission directly from the copyright holder.

Chapter 2
Machine Learning for CSI-Based Localization

Abstract Machine learning-based CSI localization techniques is important for enhancing the accuracy, adaptability, and scalability of indoor positioning systems. Building upon the foundational discussions in Chap. 1, which covered the three areas of this book. This chapter introduces the machine learning algorithms that drive these advancements forward. We explore both supervised and unsupervised learning methods, focusing on representative algorithms such as KNN, SVM, Decision Trees, k-Means, CNN, and RNN. Each technique is examined in terms of its theoretical foundation, practical strengths, and applicability to CSI-based localization. Optimization strategies, including feature selection and parameter tuning, are also discussed to improve overall system performance. A comparative analysis is presented to evaluate algorithmic effectiveness under realistic deployment conditions, addressing several challenges raised in the previous chapter.

By integrating these machine learning approaches, we also provide both theoretical insights and practical guidance for developing efficient, robust, and adaptive CSI localization systems. It builds on the basic concepts introduced in Chap. 1 and progresses toward advanced algorithmic solutions, setting the stage for subsequent chapters on offline data collection, intelligent updates, and accurate localization. This chapter is crucial for researchers and practitioners aiming to leverage machine learning for dynamic indoor environments.

Keywords Machine learning · Supervised learning · Unsupervised learning · Adaptive localization

2.1 Overview of Machine Learning

2.1.1 Basics of Machine Learning: Concepts and Definitions

Machine learning is a subset of artificial intelligence that enables systems to learn patterns from data and improve their performance over time without explicit programming (Jordan and Mitchell 2015; Zhou 2021). It relies on mathematical and statistical techniques to analyze data, uncover hidden patterns, and make predictions or decisions. The primary goal of machine learning is to generalize effectively to unseen data, ensuring robust performance in real-world applications. With the exponential growth of data and computational power, machine learning has become indispensable in fields such as image recognition, natural language processing, and financial forecasting. However, its success heavily depends on the quality and quantity of data, and it often struggles with noisy or outlier data points.

ML can be divided into three main types: supervised learning, unsupervised learning, and reinforcement learning. Supervised learning relies on labeled data to teach the model the relationship between inputs and outputs. The goal is to predict the correct output for new, unseen data. For example, in image classification, the model learns from a large set of labeled images and then uses that knowledge to classify new images. Mathematically, supervised learning tries to find a function $f(x)$ that matches the output y for a given input x. Common ways to measure errors in supervised learning include mean squared error for predicting numbers and cross-entropy for classifying things. Unlike supervised learning, unsupervised learning operates on unlabeled data, focusing instead on discovering hidden patterns or structures within the data. It works with data that doesn't have labels. It tries to find hidden patterns or groupings in the data. For example, clustering algorithms like K-means group similar data points together, while methods like PCA reduce the number of variables in the data while keeping the important information. Mathematically, clustering aims to divide a dataset $\{x_1, x_2, \ldots, x_n\}$ into k groups, where data points in the same group are very similar, and data points in different groups are very different. Reinforcement learning is different from the other two. It involves an agent learning by interacting with an environment (Szepesvári 2022). The agent takes actions, receives rewards or penalties, and tries to maximize its total reward over time. This type of learning is often used in robotics, games, and managing resources. Mathematically, the goal is to maximize the cumulative reward $R = \sum_{t=0}^{\infty} \gamma^t r_t$, where γ is a discount factor and r_t is the reward at time t.

For the CSI-based localization, machine learning can use the detailed information in CSI to estimate a device's location. It includes features like signal strength, phase, and delay, which are influenced by the environment and the positions of the devices. By training models on data collected from different locations, we can create a mapping between signal characteristics and spatial coordinates. For example, supervised learning methods like support vector machines or neural networks can use labeled input to predict the position of unknown devices. Furthermore, unsupervised learning methods such as K-means or Gaussian mixture models can group unlabeled samples to identify patterns in how signal features vary across different locations.

2.1.2 Relevance of Machine Learning in CSI Localization

ML is particularly valuable for CSI-based localization because of its ability to manage the complexity and variability of indoor wireless environments (Burghal et al. 2020; Wang et al. 2016). Traditional localization approaches, which are typically based on geometric or analytical models of signal propagation, often struggle to cope with the unpredictable behavior of wireless signals in practical scenarios. Factors such as multipath effects, NLoS propagation, and dynamic environmental changes can significantly degrade the performance of these conventional methods. Moreover, ML algorithms can learn the complex and often nonlinear relationships between CSI measurements and physical locations directly from data, without requiring explicit assumptions about the signal environment.

CSI data is inherently high-dimensional and structurally rich, containing detailed information such as amplitude and phase across multiple subcarriers and antennas. This complexity presents both challenges and opportunities for localization. Machine learning techniques, particularly those in deep learning, are highly effective at capturing patterns in such data. For instance, Convolutional Neural Networks (CNN) are capable of extracting spatial features from CSI matrices, enabling the differentiation of signal characteristics across different locations. Recurrent Neural Networks (RNN), including Long Short-Term Memory (LSTM) models, are effective in modeling temporal dependencies within sequential CSI measurements, making them suitable for scenarios involving user mobility (Hu et al. 2024).

In addition to handling the complexity of CSI data, machine learning also offers scalability in processing large volumes of data collected in modern wireless systems. As wireless infrastructure continues to expand and generate increasingly dense datasets, the ability of ML models to process and generalize from large-scale data becomes essential. Unsupervised learning techniques, such as K-means clustering and Gaussian Mixture Models, can be used to group similar CSI patterns, helping to identify spatial structures and distinguish different areas within the environment. Dimensionality reduction methods like PCA or autoencoders can simplify the data while retaining its most informative components, thus reducing computational burden and improving model efficiency.

Another important advantage of machine learning is its adaptability. Models can be updated or retrained as the environment evolves, ensuring sustained localization performance even in the presence of changes such as furniture rearrangement or varying human presence. This capability enhances the robustness, generalization, and practicality of CSI-based localization systems in real-world applications. Therefore, the combination of machine learning into CSI localization frameworks significantly improves positioning accuracy, adaptability to environmental variation, and scalability across diverse deployment conditions.

2.2 Supervised Learning Methods

2.2.1 KNN

KNN is a simple yet effective algorithm used for classification and regression tasks, as illustrated in Fig. 2.1 (Guo et al. 2003). For the CSI-based localization, KNN estimates the position of a device by leveraging CSI features such as amplitude and phase. These features are typically pre-processed through normalization to mitigate variations in signal strength before being input into the algorithm. In the training phase, a database of labeled CSI measurements collected at known locations is constructed, forming the reference set for future predictions. When a new CSI sample is received, KNN compares it against this reference set to determine the most similar entries and infer the corresponding location. Despite its simplicity, KNN is capable of delivering competitive localization performance in static or moderately dynamic environments, making it a useful baseline for evaluating more complex algorithms.

The algorithm operates by identifying the k nearest neighbors to a test sample using a distance metric. For classification tasks, it employs majority voting among the k neighbors to predict the label, whereas for regression tasks, it averages the neighbors' values. The Euclidean distance is the most commonly used metric, defined as

$$d(x_i, x_j) = \sqrt{\sum_{m=1}^{M}(x_{i,m} - x_{j,m})^2}, \tag{2.1}$$

where M represents the number of features. However, in high-dimensional CSI data, Euclidean distance can suffer from the curse of dimensionality, resulting in less reliable outcomes. In such scenarios, alternative metrics like Manhattan distance or

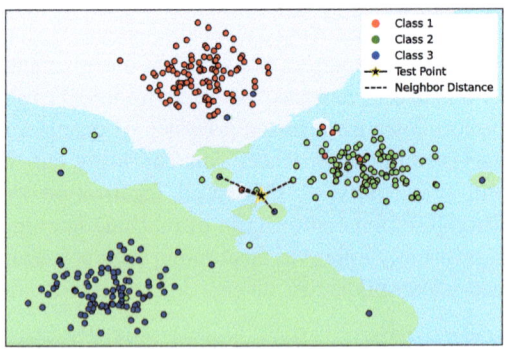

Fig. 2.1 The sructure of KNN

2.2 Supervised Learning Methods

cosine similarity are often employed, which can be described by

$$d_M(x_i, x_j) = \sqrt{(x_i - x_j)^T \Sigma^{-1}(x_i - x_j)}, \tag{2.2}$$

where Σ is the covariance matrix. And Dynamic Time Warping (DTW) can improve results by accounting for feature correlations and temporal dynamics, respectively.

The performance of KNN in localization tasks is highly influenced by the number of neighbors considered. Using too few may result in overfitting, as the prediction becomes overly sensitive to local noise, while too many can lead to underfitting by oversmoothing the estimation. Therefore, determining an appropriate value is critical and is typically accomplished through cross-validation. The optimal choice also depends on the spatial distribution of the data: fewer neighbors tend to perform better in densely sampled areas, whereas a larger neighborhood is more suitable for sparsely populated regions. With careful parameter tuning and proper preprocessing, KNN can offer reliable and accurate localization results.

2.2.2 SVM

Support Vector Machine (SVM) is a robust supervised learning method known for its ability to handle both linear and nonlinear classification and regression tasks (Noble 2006). By focusing on minimizing structural risk, SVM is particularly effective in high-dimensional spaces and with small datasets, offering strong generalization capabilities. However, its performance can be hindered when dealing with large volumes of data, and the choice of kernel function is important for determining the model's effectiveness. Commonly used kernels include linear, polynomial, and radial basis function (RBF), with the latter offering the most flexibility and widespread application due to its adaptability to complex decision boundaries.

The core idea of SVM is to identify an optimal hyperplane that separates data points from different classes. In the case of linearly separable data, this separating boundary is chosen to maximize the margin, which refers to the smallest distance between the boundary and the nearest training samples known as support vectors. These support vectors are the only data points that influence the final model. For binary classification, the decision function is

$$f(x) = \text{sign}(w^T x + b), \tag{2.3}$$

where w is the weight vector, and b is the bias term. In order to maximize the classification margin, the optimization problem is formulated as minimizing the norm of the weight vector w, which is equivalent to solving $\min_{w,b} \frac{1}{2}\|w\|^2$. Subject to $y_i(w^T x_i + b) \geq 1$ for all samples i. This optimization problem can be efficiently solved using the method of Lagrange multipliers, which is suitable for handling the constraints in this problem. For non-linearly separable data, SVM uses kernel functions to map data into a higher-dimensional space where a linear hyperplane can be found. The RBF kernel is defined as

$$K(x_i, x_j) = \exp\left(-\gamma \|x_i - x_j\|^2\right), \tag{2.4}$$

where γ controls the kernel's flexibility. The RBF kernel must satisfy the Mercer condition to ensure the kernel matrix is positive definite.

The effectiveness of SVM in CSI localization largely depends on how the hyperplane is optimized and which kernel function is selected, as these factors influence the model's ability to distinguish CSI features from different spatial positions (Zhou et al. 2017). SVM can be applied to classify signal patterns into specific location areas or to directly estimate position coordinates, making it well-suited for both indoor and multi-target localization scenarios. Given the complex and high-dimensional nature of CSI data, kernel functions are used to transform the input space, allowing for better class separation and more accurate localization performance.

2.2.3 Decision Trees and Random Forests

Decision trees offer a structured and intuitive approach to learning patterns from data, making them a popular choice for both classification and regression tasks (Kotsiantis 2013). They function by recursively partitioning the feature space based on feature values, aiming to group samples with similar target outputs. This process relies on metrics such as Gini impurity or information gain to guide the splits. While decision trees are easy to interpret and implement, they are also vulnerable to overfitting, especially when the model becomes overly complex or when noise exists in the training data.

It includes internal nodes, branches, and leaf nodes. Each internal node evaluates a specific feature, the branches represent possible outcomes of that evaluation, and the leaf nodes provide a class label in classification tasks or a predicted value in regression tasks. In classification, the assigned label is determined by the majority of training samples reaching the leaf node, while in regression, the prediction is typically the average of these values. Constructing a decision tree involves identifying the most informative feature at each node to split the data. Information gain evaluates how much a feature reduces uncertainty in the dataset, thereby improving the model's predictive performance by

$$\text{IG}(D, f) = H(D) - \sum_{v \in f} \frac{|D_v|}{|D|} H(D_v), \tag{2.5}$$

2.2 Supervised Learning Methods

where $H(D)$ is the entropy of dataset D, and D_v is the subset in which feature f takes value v. And Gini impurity also reflects the probability of misclassification within a subset as

$$\text{Gini}(D) = 1 - \sum_{k=1}^{K} p_k^2, \tag{2.6}$$

where p_k denotes the proportion of class k in D. When applied to CSI localization, the tree may initially split the data based on normalized signal strength from a specific subcarrier. Subsequent nodes then evaluate additional CSI features, such as phase or latency, refining the classification until a stopping criterion is met–such as reaching a maximum depth or encountering a node with too few samples. Since decision trees are prone to overfitting and sensitive to noise, ensuring feature quality and reducing redundancy are essential to achieving stable localization results.

Random forests extend decision trees by using an ensemble learning approach that significantly enhances robustness and generalization (Biau 2012). Rather than relying on a single tree, they aggregate the predictions of multiple trees, each trained on a bootstrapped sample of the original data. Moreover, by selecting a random subset of features at each split, the model introduces further variability, which reduces the correlation between trees and mitigates the risk of overfitting. This ensemble strategy is particularly beneficial in scenarios where the input data is complex and subject to environmental fluctuations, as is often the case in CSI-based localization. The diversity among trees allows the model to capture subtle, location-dependent patterns in the CSI features while maintaining resistance to noise and outliers.

2.2.4 Neural Networks

Neural networks are well known for their ability to model complex nonlinear patterns (Abiodun et al. 2018). Deep architectures have achieved impressive performance in fields such as image classification, speech recognition, and indoor localization. These models learn from large-scale data and adapt to diverse input features through layered transformations. A typical neural network consists of an input layer, multiple hidden layers, and an output layer. Each neuron computes a weighted sum of its inputs, adds a bias term, and passes the result through an activation function. The depth of the network is determined by the number of hidden layers, which enables the model to extract increasingly abstract representations of the input. The output of the lth layer is given by

$$a^{(l)} = f^{(l)}(W^{(l)} a^{(l-1)} + b^{(l)}), \tag{2.7}$$

where $a^{(l)}$ denotes the activation vector, $W^{(l)}$ the weight matrix, $b^{(l)}$ the bias vector, and $f^{(l)}$ the activation function. Common activation functions include sigmoid, hyperbolic tangent, and ReLU.

In CSI-based localization, neural networks learn the nonlinear relationship between CSI features and location coordinates (Wang et al. 2016). The input layer takes CSI features, and the output layer predicts location coordinates. Hidden layers automatically learn high-level features from raw CSI data, capturing the complex signal-location relationship. Training involves iteratively adjusting weights and biases to minimize the difference between predicted and actual outputs. Optimization is typically performed using gradient-based algorithms, such as Stochastic Gradient Descent (SGD) or its variants (e.g., Adam, RMSProp). For regression tasks, the mean squared error (MSE) serves as the loss function, while cross-entropy loss is commonly used for classification tasks to measure the divergence between predicted and true distributions.

Different neural architectures offer distinct strengths in processing CSI data. CNN is effective at capturing spatial features, especially when CSI is transformed into image-like representations. RNN and LSTM networks are suited for sequential modeling, making them ideal for tracking location changes over time. Transformer-based models, originally developed for language tasks, have recently shown strong potential in localization due to their ability to model long-range dependencies without relying on recurrence. By using attention mechanisms, Transformers can capture global context within CSI sequences, improving localization performance in dynamic and cluttered environments.

2.3 Unsupervised Learning Methods

2.3.1 K-Means Clustering

K-Means clustering is an unsupervised learning algorithm that partitions a dataset into k clusters, as shown in Fig. 2.2 (Kodinariya et al. 2013). The algorithm assigns each data point to the nearest cluster center and updates the centers based on the mean of the assigned points. Its goal is to minimize the variance within each cluster. One limitation of K-Means is its sensitivity to the initial placement of cluster centers, which may lead to convergence at a local optimum.

The algorithm begins by randomly selecting k cluster centers. Methods such as K-Means++ are often used to improve the initialization. For each data point, the algorithm calculates the distance to every cluster center, typically using the Euclidean metric, and assigns the point to the closest one. After all assignments are made, the cluster centers are updated by averaging the positions of the points in each cluster. This procedure repeats until a convergence condition is met, such as reaching a maximum number of iterations or observing minimal change in center locations. The objective function minimizes the total within-cluster sum of squared

Fig. 2.2 The structure of K-means clustering

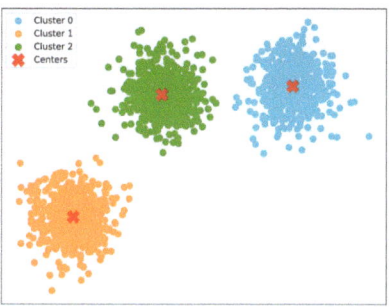

distances as

$$d = \min_{\mu_1,\ldots,\mu_k} \sum_{i=1}^{N} \sum_{j=1}^{k} r_{ij} |x_i - \mu_j|^2, \qquad (2.8)$$

where r_{ij} equals 1 if data point x_i is assigned to cluster j, and 0 otherwise. The variable μ_j denotes the center of cluster j. While K-Means is straightforward and efficient, it is affected by noise and outliers, which can significantly alter the clustering results.

In CSI-based localization, K-Means can help reveal structural similarities in the feature space across various spatial regions. With a suitable choice of k, determined by approaches such as the elbow method or silhouette analysis, the algorithm groups CSI measurements into clusters that correspond to different physical locations. The center of each cluster provides a representative CSI profile for that area. Prior to clustering, applying normalization to the CSI data helps avoid skewed results caused by differences in feature magnitudes.

2.3.2 DBSCAN

DBSCAN (Density-Based Spatial Clustering of Applications with Noise) is a clustering algorithm designed to discover clusters of arbitrary shapes and identify noise in spatial datasets, as shown in Fig. 2.3 (Schubert et al. 2017; Khan et al. 2014). Unlike partition-based algorithms that assume spherical cluster structures, DBSCAN groups data points based on local point density. A cluster is defined as a set of points that are closely packed together, where density is measured by the number of neighboring points within a given radius. A point is considered a core point if its neighborhood contains at least a specified minimum number of other points; otherwise, it may be classified as a border point or noise.

The algorithm proceeds by scanning the dataset to find core points and expanding clusters from them. Starting with an unvisited point, DBSCAN checks whether the number of points within its ϵ-neighborhood meets the minimum requirement. If

Fig. 2.3 The structure of DBSCAN

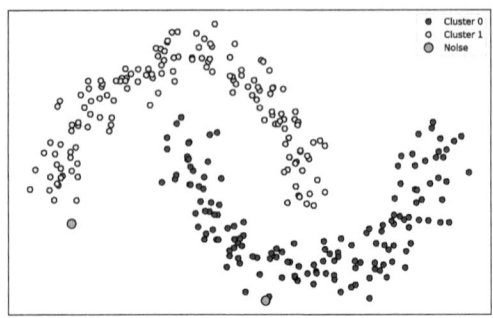

so, the point becomes the seed of a new cluster, and the cluster grows by recursively including all density-reachable points. Points that are not reachable from any cluster are labeled as noise. Two points x_i and x_j are density-connected if there exists a sequence of intermediate points between them, such that each consecutive pair lies within ϵ distance and satisfies the core point condition. This density-based notion allows DBSCAN to flexibly adapt to the shape and scale of the data distribution.

When applied to CSI-based localization, DBSCAN provides a non-parametric approach for uncovering spatial structures in CSI feature space. It can group similar CSI patterns that correspond to specific physical regions, even when clusters have irregular boundaries. Moreover, DBSCAN can automatically detect and isolate outliers caused by environmental dynamics, device interference, or sensor noise. Removing these outliers improves the robustness of localization models. Compared to methods that require prior knowledge of the number of clusters, DBSCAN offers the advantage of adaptive discovery, which is particularly useful in real-world deployments where the number of location zones is not fixed.

2.3.3 WPCA for Dimensionality Reduction

Principal Component Analysis (PCA) is a classical dimensionality reduction technique that transforms high-dimensional data into a lower-dimensional subspace while preserving as much variance as possible (Jolliffe and Cadima 2016). Given a dataset $X = x_1, x_2, \ldots, x_N$, where each $x_i \in \mathbb{R}^D$, PCA first computes the covariance matrix

$$C = \frac{1}{N} \sum_{i=1}^{N} (x_i - \mu)(x_i - \mu)^T, \tag{2.9}$$

where the mean vector is $\mu = \frac{1}{N} \sum_{i=1}^{N} x_i$. The principal components are obtained by performing eigenvalue decomposition on C, where the eigenvectors correspond to directions of maximum variance. By projecting data onto the top eigenvec-

2.3 Unsupervised Learning Methods

tors, PCA retains the most informative dimensions while reducing noise and computational complexity. This method is widely used for feature compression, visualization, and as a preprocessing step in machine learning pipelines.

However, traditional PCA assumes equal importance for all samples and dimensions, which makes it sensitive to noise and outliers. This limitation is particularly evident in CSI-based localization tasks, where wireless signals are affected by multipath propagation, interference, and device imperfections. These distortions can skew the principal components and degrade the performance of downstream localization models. To address this, Weighted PCA (WPCA) introduces a weighting mechanism that assigns different importance to each data point. For a dataset with associated weights w_i, the weighted covariance matrix is defined as

$$C_w = \frac{1}{\sum_{i=1}^{N} w_i} \sum_{i=1}^{N} w_i (x_i - \mu_w)(x_i - \mu_w)^T, \qquad (2.10)$$

where the weighted mean is given by $\mu_w = \frac{\sum_{i=1}^{N} w_i x_i}{\sum_{i=1}^{N} w_i}$. The eigenvectors of C_w define the weighted principal components. It can make the model emphasize more reliable or informative samples while reducing the influence of noisy or corrupted data, resulting in improved robustness.

In CSI-based localization, the input features typically include both amplitude and phase information across multiple subcarriers. Amplitude features tend to be more stable and location-dependent, whereas phase features are susceptible to noise from hardware imperfections such as carrier frequency offset and sampling asynchrony. WPCA enables an adaptive weighting strategy where higher weights are assigned to more stable amplitude components, and lower weights are applied to noisy phase components. This selective emphasis improves the signal-to-noise ratio of the transformed features, enhancing localization accuracy. Compared to standard PCA, which treats all CSI components uniformly, WPCA leverages domain knowledge to suppress irrelevant variations and preserve discriminative structure in the data. By adaptively weighting CSI features based on their reliability and informativeness, WPCA produces a compact and noise-resilient representation that improves the performance of learning-based localization systems.

2.3.4 t-SNE

t-SNE (t-distributed stochastic neighbor embedding) is a nonlinear dimensionality reduction technique designed for visualizing high-dimensional data in two or three dimensions, as illustrated in Fig. 2.4 (Maaten and Hinton 2008). It focuses on preserving local structures by modeling pairwise similarities between data points. Despite its effectiveness in visual exploration, t-SNE is computationally expensive and highly sensitive to hyperparameters such as perplexity, which must be carefully tuned for optimal performance.

Fig. 2.4 The structure of t-SNE

The algorithm begins by computing the similarity between two high-dimensional data points x_i and x_j using a conditional probability distribution by

$$p_{j|i} = \frac{\exp(-\|x_i - x_j\|^2/2\sigma_i^2)}{\sum_{k \neq i} \exp(-\|x_i - x_k\|^2/2\sigma_i^2)}, \quad (2.11)$$

where σ_i controls the local scale of neighborhood around x_i. To ensure symmetry, the joint probability is then defined as $p_{ij} = \frac{p_{j|i} + p_{i|j}}{2N}$. In the low-dimensional embedding, similarities between points y_i and y_j are measured using a Student's t-distribution with a single degree of freedom (i.e., $\alpha = 1$), which helps mitigate the "crowding problem." The corresponding probability q_{ij} is computed as

$$q_{ij} = \frac{(1 + \|y_i - y_j\|^2)^{-1}}{\sum_{k \neq l}(1 + \|y_k - y_l\|^2)^{-1}}. \quad (2.12)$$

The objective of t-SNE is to make the low-dimensional distribution $Q = q_{ij}$ as close as possible to the high-dimensional distribution $P = p_{ij}$ by minimizing the Kullback-Leibler divergence as

$$\mathrm{KL}(P \| Q) = \sum_{i \neq j} p_{ij} \log \frac{p_{ij}}{q_{ij}}. \quad (2.13)$$

Gradient descent is used to optimize the embedding, with careful adjustment of momentum and learning rate to accelerate convergence and improve stability. While t-SNE excels at uncovering complex structures such as clusters and manifolds, it is typically used for visualization rather than downstream tasks due to its computational cost and lack of scalability.

For CSI-based localization, t-SNE offers an effective approach to visualize high-dimensional CSI features by projecting them into a lower-dimensional space, typically 2D or 3D. By preserving local relationships between data points, t-SNE helps identify patterns and clusters that correspond to different physical locations. Well-separated clusters in the low-dimensional space often indicate that CSI features contain location-specific information. Such visualizations can provide insights into the quality and discriminability of the features, helping to assess whether the data can distinguish different locations effectively. Although t-SNE is not directly used for prediction, it is valuable for evaluating feature quality, detecting anomalies, and guiding model design. Proper tuning of parameters such as perplexity and learning rate is essential to ensure meaningful embeddings.

2.4 Deep Learning Models

2.4.1 CNN for Spatial Feature Extraction

CNN is deep learning architectures particularly suitable for processing grid-structured data, such as images (Li et al. 2021). Their ability to extract local patterns makes them a strong candidate for CSI-based localization tasks, where spatial relationships within the signal can reveal information about a user's position. However, CNNs typically require large datasets and substantial computational resources. Moreover, they are often combined with sequence-processing models when applied to time-dependent data.

A standard CNN includes three main components that work together to extract and process spatial features. The first is the convolutional layer, which uses multiple learnable filters to scan the input tensor. Each filter slides across the data with a fixed stride and computes feature maps by detecting local patterns. This process can be represented as

$$z_{i,j} = \sum_{m=1}^{M}\sum_{n=1}^{N} x_{i+m,j+n} \cdot w_{m,n} + b, \qquad (2.14)$$

where x denotes the input, w the filter weights, and b the bias term. Next, pooling layers reduce the spatial resolution of feature maps while preserving important information. Max pooling is commonly used, calculated by

$$z_{pool} = \max(z_{i:i+k, j:j+k}), \qquad (2.15)$$

where k is the size of the pooling window. This operation not only reduces computation but also improves robustness to slight variations in the input. Fully connected layers follow the convolutional and pooling stages to integrate the spatial features and perform location prediction. Regularization methods such as Dropout are applied during training to reduce overfitting. The output of a Dropout layer can be described by

$$y = \text{Dropout}(Wx + b, p_{drop}), \qquad (2.16)$$

where p_{drop} is the dropout rate, and W and b are learnable parameters.

When applied to CSI data, CNN can treat amplitude and phase components as separate input channels, similar to the way color channels are handled in image data (Wang et al. 2018). To ensure consistent scaling across inputs, normalization or standardization is typically performed before training. With properly designed filter sizes and network depth, CNN can learn spatial features that reflect multipath effects and channel changes caused by user movement or environmental obstacles. During training, model parameters are updated using optimization methods such as stochastic gradient descent or Adam. The loss function is selected based on the task, where mean squared error is commonly used for regression and cross-entropy is used for classification. Regularization methods like L2 weight decay are also applied to improve the model's ability to generalize.

2.4.2 RNN and LSTM

RNN is well-suited for processing sequential data, as it maintains a hidden state that evolves over time, allowing the model to retain and utilize past information. However, traditional RNNs face challenges in learning long-term dependencies due to the vanishing gradient problem during backpropagation. To overcome this, LSTM networks introduce a more structured mechanism that enables more stable gradient flow over long sequences (Hochreiter and Schmidhuber 1997).

LSTM units incorporate a memory cell along with three types of gates: the forget gate, input gate, and output gate, as shown in Fig. 2.5. These components work together to control how information is stored, updated, and propagated through time. At each time step t, the LSTM takes the previous hidden state h_{t-1} and the current input x_t, and performs the following computations.

First, the forget gate determines which information from the previous cell state C_{t-1} should be discarded $f_t = \sigma_f(W_f[h_{t-1}, x_t] + b_f)$, where W_f is the weight matrix, b_f is the bias, and σ_f is usually a sigmoid activation function. Next, the input gate decides what new information will be added to the cell state:

$$i_t = \sigma_i(W_i[h_{t-1}, x_t] + b_i), \qquad (2.17)$$

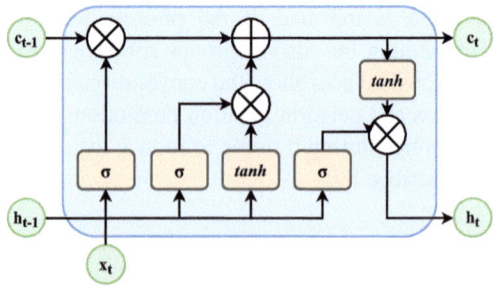

Fig. 2.5 The structure of LSTM

2.4 Deep Learning Models

where σ_i is typically a sigmoid function, and W_i, b_i are trainable parameters. Alongside the input gate, the candidate cell state is computed to represent the new content to be added $\tilde{C}_t = \sigma_c(W_c[h_{t-1}, x_t] + b_c)$, where σ_c is generally a tanh activation function. The new cell state is then updated by combining the retained previous memory and the candidate state:

$$C_t = f_t \odot C_{t-1} + i_t \odot \tilde{C}_t, \tag{2.18}$$

where $\odot$ denotes element-wise multiplication. The output gate determines which parts of the cell state contribute to the hidden state:

$$o_t = \sigma_o(W_o[h_{t-1}, x_t] + b_o), \tag{2.19}$$

and the hidden state is updated as $h_t = o_t \odot \sigma_h(C_t)$, where σ_h is usually another tanh function. This gating mechanism enables LSTM to selectively remember or forget information over long sequences, effectively modeling dependencies over time.

In CSI-based localization, time-series data reflects variations in the wireless channel as the user moves or the environment changes (Zhang et al. 2020). LSTM networks are particularly effective in capturing these temporal dynamics. By learning how CSI values evolve, the model can predict motion trends, track location over time, and filter out transient noise. Therefore, it makes LSTM suitable for tasks such as trajectory estimation, motion classification, and real-time tracking.

2.4.3 Autoencoders

Autoencoders are neural networks designed for unsupervised learning, as shown in Fig. 2.6 (Li et al. 2023). They learn to compress input data into a compact representation and then reconstruct it from this latent form. An autoencoder consists of two components: an encoder and a decoder. The encoder maps the input x to a lower-dimensional latent representation h:

$$h = f(x) = \sigma(Wx + b), \tag{2.20}$$

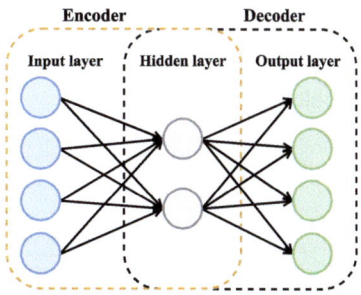

Fig. 2.6 The structure of autoencoders

where W is the weight matrix, b is the bias vector, and σ is an activation function such as ReLU or Sigmoid. This step captures key patterns in the input while reducing dimensionality. The decoder reconstructs the input from the latent representation:

$$\hat{x} = g(h) = \sigma'(W'h + b'), \tag{2.21}$$

where W' and b' are the decoder's weight matrix and bias, and σ' is typically set to match σ for simplicity. The goal during training is to minimize the reconstruction error, thereby ensuring that the latent representation retains essential information from the original input, and often using mean squared error as the loss function.

For CSI-based localization, autoencoders can reduce the dimensionality of high-resolution CSI data while preserving location-related features. CSI often consists of amplitude and phase information across multiple subcarriers, forming high-dimensional input vectors. These vectors can be fed into the encoder to extract a compact representation that captures spatial characteristics of the wireless environment. The latent features obtained from the encoder can reflect consistent patterns associated with specific indoor locations. These features may include unique variations in amplitude and phase that result from multipath propagation or user movement. Once learned, the latent representations can be used as inputs for downstream localization models, such as classifiers or regression networks, to estimate user positions more efficiently and robustly.

2.4.4 GAN

2.4.5 Generative Adversarial Networks for Data Augmentation

Generative Adversarial Networks (GANs) consist of two neural networks: a generator and a discriminator (Creswell et al. 2018; Saxena and Cao 2021). These two components are trained together in an adversarial setting, as shown in Fig. 2.7. The generator is responsible for producing synthetic data samples, while the discriminator attempts to distinguish between real and generated data. During training, the generator gradually improves its ability to produce realistic outputs, and the discriminator becomes more adept at identifying differences between real and fake samples.

The training objective of GANs can be formulated as a minimax optimization problem:

$$\min_{G} \max_{D} V(D, G) = \mathbb{E}_{x \sim p_{\text{data}}(x)}[\log D(x)] + \mathbb{E}_{z \sim p_z(z)}[\log(1 - D(G(z)))], \tag{2.22}$$

2.4 Deep Learning Models

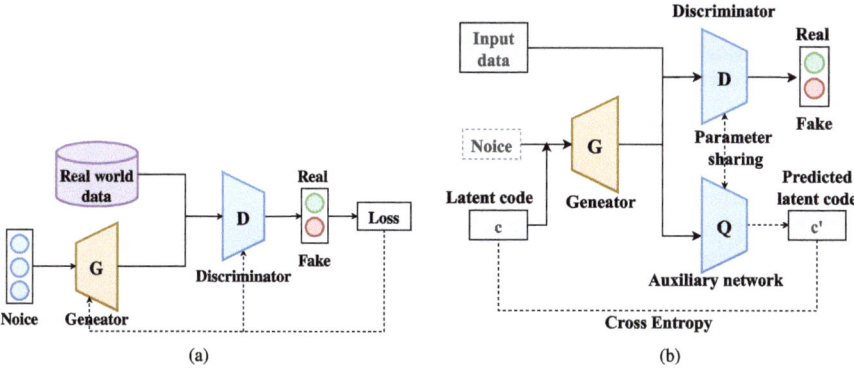

Fig. 2.7 The different structures of GANs. (**a**) Original GAN. (**b**) InfoGAN

where G and D denote the generator and discriminator, respectively. The term $p_{\text{data}}(x)$ represents the real data distribution, while $p_z(z)$ is the noise distribution used as input to the generator. The first expectation term encourages the discriminator to assign high probabilities to real data samples. The second term penalizes it for incorrectly classifying generated data as real.

In practice, these expectations are estimated using mini-batch samples. Given a batch of real data $\{x_1, x_2, \ldots, x_N\}$, the first term can be approximated as

$$\frac{1}{N} \sum_{i=1}^{N} \log D(x_i), \tag{2.23}$$

and for a batch of noise vectors $\{z_1, z_2, \ldots, z_M\}$, the second term becomes

$$\frac{1}{M} \sum_{j=1}^{M} \log(1 - D(G(z_j))). \tag{2.24}$$

The generator is trained to minimize this objective by producing data that the discriminator is less capable of distinguishing from real samples, while the discriminator seeks to maximize it by improving its classification accuracy.

In CSI-based localization, the scarcity of labeled data often limits model performance, particularly in complex or resource-constrained environments (Wang et al. 2021). GANs provide an effective strategy to address this issue by generating synthetic CSI data that reflects the statistical characteristics of real measurements. This is especially advantageous in scenarios such as industrial settings, where electromagnetic interference and structural occlusions complicate large-scale data collection. By learning from real CSI distributions, the generator captures key spatial features embedded in amplitude and phase variations caused by multipath propagation. The synthesized data can be used to augment training sets, improving the robustness and generalization ability of localization models. In addition, the synthetic samples facilitate the analysis of environment-specific propagation patterns, supporting model adaptation to diverse deployment conditions.

2.5 Challenges and Opportunities

The comparison of representative ML algorithms used in CSI-based localization is listed in Table 2.1. ML offers powerful tools for extracting features and building accurate models, yet several challenges remain in terms of data collection efficiency, real-time adaptability, and localization accuracy.

Model performance is strongly influenced by the quality and quantity of training data. However, collecting large-scale, labeled CSI datasets in dynamic environments remains labor-intensive and time-consuming. Conventional fingerprinting methods often fail to adapt to environmental changes, such as user mobility, structural modifications, or varying interference levels, leading to a decline in localization accuracy. This underscores the need for adaptive frameworks that support online updates and real-time learning. Another major challenge lies in achieving high-precision localization in complex indoor environments, which often requires models capable of managing multipath propagation, signal fluctuation, and hardware heterogeneity. These requirements typically lead to increased computational burden, making real-time inference difficult on resource-constrained devices. Striking a balance between inference speed, accuracy, and robustness remains an open problem.

Despite these challenges, several promising directions offer potential solutions. Techniques such as lightweight deep networks and knowledge distillation can reduce computational load without significantly sacrificing accuracy. Multi-source data fusion, combining information from CSI, inertial sensors, or vision, can improve resilience and accuracy. Moreover, transfer learning and reinforcement learning enable models to adapt efficiently to new environments with minimal retraining. Federated learning and self-supervised approaches help address privacy and data scarcity concerns by enabling decentralized training with unlabeled data. Generative models like GANs further contribute by synthesizing realistic training data under various channel conditions. Collectively, these approaches open new opportunities for developing robust, scalable, and intelligent localization systems that align with the demands of IoT and ISAC applications.

2.5 Challenges and Opportunities

Table 2.1 Comparison of machine learning algorithms for CSI localization

Algorithm type	Method	Applicable scenarios	Advantages	Limitations
Supervised learning	KNN	Small-scale indoor environments with limited dynamic changes	Easy to implement, non-parametric, and performs well with dense labeled data	Computationally expensive during inference, performance degrades in high-dimensional spaces
	SVM	High-dimensional CSI data, scenarios with limited training samples	Good generalization ability and effective handling of nonlinear boundaries	Sensitive to parameter tuning, high training complexity
	Decision Tree	Classification and regression tasks in moderately complex environments	Intuitive model structure, interpretable decision process	Tends to overfit, sensitive to noisy data
	Neural Network (Fully Connected)	Scenarios requiring modeling of complex feature interactions	Capable of learning nonlinear mappings, scalable to large datasets	Requires extensive training data, lacks interpretability
Unsupervised learning	K-means clustering	Exploratory analysis and unsupervised grouping of CSI patterns	Simple to use, fast convergence, interpretable results	Sensitive to initialization and outliers, assumes spherical clusters
	DBSCAN	Environments with noise and irregular cluster shapes	Handles arbitrary-shaped clusters, robust to noise	Difficult to select optimal parameters, performance affected by density variations
	WPCA	Dimensionality reduction and signal denoising for high-dimensional CSI	Preserves variance structure, improves feature interpretability	Sensitive to noise and scaling, relies on proper weighting schemes
	t-SNE	Visualization and structure discovery in high-dimensional CSI datasets	Preserves local relationships, useful for understanding latent structures	Computationally intensive, not suitable for large-scale real-time applications

(continued)

Table 2.1 (continued)

Algorithm type	Method	Applicable scenarios	Advantages	Limitations
Deep learning	CNN	Spatial pattern extraction from CSI amplitude and phase maps	Strong ability to extract local features and spatial correlations	Requires large datasets, computationally demanding
	RNN	Time-series modeling of CSI variations due to motion	Captures temporal dependencies effectively	Prone to vanishing gradients, limited in long sequences
	Autoencoder	Feature compression and unsupervised representation learning	Reduces dimensionality while retaining key information	Sensitive to data noise and distribution shifts
	GAN	Data augmentation and synthetic CSI generation	Enhances training with diverse samples, captures complex data distributions	Unstable training dynamics, risk of mode collapse

References

Abiodun OI, Jantan A, Omolara AE, Dada KV, Mohamed NA, Arshad H (2018) State-of-the-art in artificial neural network applications: a survey. Heliyon 4(11):e00938

Biau G (2012) Analysis of a random forests model. J Mach Learn Res 13:1063–1095

Burghal D, Ravi AT, Rao V, Alghafis AA, Molisch AF (2020) A comprehensive survey of machine learning based localization with wireless signals. arXiv:201211171

Creswell A, White T, Dumoulin V, Arulkumaran K, Sengupta B, Bharath AA (2018) Generative adversarial networks: an overview. IEEE Signal Process Mag 35(1):53–65

Guo G, Wang H, Bell D, Bi Y, Greer K (2003) KNN model-based approach in classification. In: On the move to meaningful internet systems 2003: CoopIS, DOA, and ODBASE: OTM confederated international conferences, CoopIS, DOA, and ODBASE 2003, Catania, Sicily, Italy, November 3–7, 2003. Proceedings, pp 986–996

Hochreiter S, Schmidhuber J (1997) Long short-term memory. Neural Comput 9(8):1735–1780. https://doi.org/10.1162/neco.1997.9.8.1735

Hu S, Zhang C, Liu J, Zhu X, Li L (2024) Proto-CSNet: a prototype network model integrating CNN and self-attention for enhanced human activity recognition. In: IEEE The 20th international conference on mobility, sensing and networking

Jolliffe IT, Cadima J (2016) Principal component analysis: a review and recent developments. Philos Trans R Soc A Math Phys Eng Sci 374(2065):20150202

Jordan MI, Mitchell TM (2015) Machine learning: trends, perspectives, and prospects. Science 349(6245):255–260

Khan K, Rehman SU, Aziz K, Fong S, Sarasvady S (2014) DBSCAN: past, present and future. In: The fifth international conference on the applications of digital information and web technologies (ICADIWT 2014), pp 232–238

Kodinariya TM, Makwana PR, et al. (2013) Review on determining number of cluster in k-means clustering. Int J 1(6):90–95

Kotsiantis SB (2013) Decision trees: a recent overview. Artif Intell Rev 39:261–283

Li P, Pei Y, Li J (2023) A comprehensive survey on design and application of autoencoder in deep learning. Appl Soft Comput 138:110176

Li Z, Liu F, Yang W, Peng S, Zhou J (2021) A survey of convolutional neural networks: analysis, applications, and prospects. IEEE Trans Neural Netw Learn Syst 33(12):6999–7019

Maaten LVD, Hinton G (2008) Visualizing data using t-SNE. J Mach Learn Res 9:2579–2605

Noble WS (2006) What is a support vector machine? Nat Biotechnol 24(12):1565–1567

Saxena D, Cao J (2021) Generative adversarial networks (GANs) challenges, solutions, and future directions. ACM Comput Surv 54(3):1–42

Schubert E, Sander J, Ester M, Kriegel HP, Xu X (2017) DBSCAN revisited, revisited: why and how you should (still) use DBSCAN. ACM Trans Database Syst 42(3):1–21

Szepesvári C (2022) Algorithms for reinforcement learning. Springer, Berlin

Wang X, Gao L, Mao S, Pandey S (2016) CSI-based fingerprinting for indoor localization: a deep learning approach. IEEE Trans Veh Technol 66(1):763–776

Wang X, Wang X, Mao S (2018) Deep convolutional neural networks for indoor localization with CSI images. IEEE Trans Netw Sci Eng 7(1):316–327

Wang D, Yang J, Cui W, Xie L, Sun S (2021) Multimodal CSI-based human activity recognition using gans. IEEE Internet Things J 8(24):17345–17355

Zhang Y, Qu C, Wang Y (2020) An indoor positioning method based on CSI by using features optimization mechanism with LSTM. IEEE Sens J 20(9):4868–4878

Zhou ZH (2021) Machine learning. Springer, Berlin

Zhou R, Lu X, Zhao P, Chen J (2017) Device-free presence detection and localization with SVM and CSI fingerprinting. IEEE Sens J 17(23):7990–7999

Open Access This chapter is licensed under the terms of the Creative Commons Attribution 4.0 International License (http://creativecommons.org/licenses/by/4.0/), which permits use, sharing, adaptation, distribution and reproduction in any medium or format, as long as you give appropriate credit to the original author(s) and the source, provide a link to the Creative Commons license and indicate if changes were made.

The images or other third party material in this chapter are included in the chapter's Creative Commons license, unless indicated otherwise in a credit line to the material. If material is not included in the chapter's Creative Commons license and your intended use is not permitted by statutory regulation or exceeds the permitted use, you will need to obtain permission directly from the copyright holder.

Chapter 3
Efficient Offline Data Collection

Abstract This chapter examines key technologies for offline data collection in ISAC systems. It first compares manual and automated data collection methods, highlighting their strengths, weaknesses, and applicable scenarios, while addressing the challenges of balancing data quality and quantity in offline CSI collection. The chapter then details the design of automated systems, including robotic devices and power-driven sampling techniques, to streamline large-scale data collection. Additionally, it contrasts device-based and device-free methods, proposing strategies to minimize data loss.

Two core algorithms are introduced: the A3C-IP algorithm, which uses asynchronous reinforcement learning to optimize data collection paths and fingerprint prediction, and the CPPU algorithm, which integrates GAN to dynamically update CSI data and improve collection efficiency through optimal path planning. Performance evaluations validate their effectiveness in intelligent positioning systems, offering robust solutions for efficient offline data collection.

Keywords Offline data collection · Reinforcement learning · Integrated sensing and communication

3.1 Overview of Offline Data Collection Techniques

3.1.1 Manual vs. Automated Data Collection Approaches

Accurate and efficient offline data collection is essential for the development of intelligent localization systems, as it directly affects the quality of training data and the reliability of algorithm evaluation. As shown in Fig. 3.1, offline data acquisition methods are generally divided into manual and automated approaches. These two methods differ in aspects such as precision, efficiency, required resources, and technical difficulty. In practice, the choice of collection strategy often depends on the specific needs of the application scenario and the desired level of positioning accuracy.

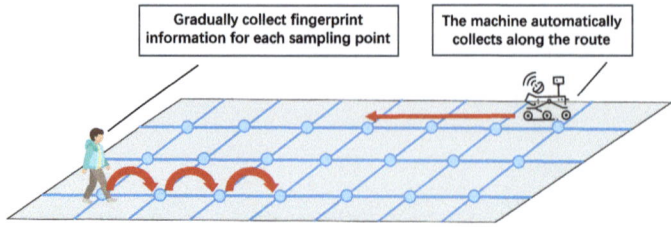

Fig. 3.1 Manual and automated data collection approaches

Manual data collection remains the most established methodology, requiring direct researcher involvement and utilizing specialized equipment including Wi-Fi CSI devices, UWB sensors, and IMUs to capture signal measurements while manually recording ground-truth positions in controlled environments. The approach primarily employs point-by-point measurement at predefined locations for high-precision positioning systems (though with efficiency limitations), trajectory tracking enhanced by RTK GPS or laser rangefinders for continuous path data (despite being labor-intensive), and camera-assisted labeling through video analysis to improve annotation accuracy (while requiring additional processing). While offering superior data quality and precise annotations crucial for research validation, this conventional method presents significant drawbacks including substantial resource requirements, operational inefficiency, and inherent dependence on human accuracy throughout the entire data acquisition process.

Automated data collection employs computer-controlled sensing technologies to enable intelligent data acquisition with minimal human intervention, thereby improving efficiency and consistency. Key methods include robot-based collection that leverages SLAM or UWB/LiDAR-equipped mobile robots for autonomous navigation and precise data capture, wearable devices that automatically record motion trajectories and environmental signals for human behavior analysis, and drone-based systems that combine GNSS and visual positioning for large-scale outdoor data collection (Macario Barros et al. 2022). In addition, edge-computing solutions using distributed IoT nodes support autonomous data processing without centralized infrastructure. Although these approaches improve scalability and reduce manual effort, they also face challenges such as high equipment costs, system complexity, and potential accuracy limitations in automated positioning techniques like SLAM.

In summary, offline data collection for intelligent positioning systems often involves both manual and automated approaches, as each offers distinct advantages. Manual data collection provides high precision and reliable data quality, which makes it particularly valuable for validation studies and the development of high-accuracy positioning models. However, this method is limited by its inefficiency and substantial resource demands. Automated data collection, supported by intelligent sensing devices and automation technologies, offers improvements in efficiency and scalability, though it may encounter limitations in positioning accuracy and requires considerable hardware investment. In practical deployment scenarios, combining

3.1 Overview of Offline Data Collection Techniques 49

manual fine-grained calibration with automated large-scale data acquisition can help maintain data quality while improving collection efficiency and system robustness, thereby delivering more dependable datasets for algorithm training and model development.

3.1.2 Key Requirements for Offline CSI Collection

In wireless signal-based positioning systems, CSI quality has a direct impact on model training effectiveness and localization accuracy (Guo et al. 2022). To ensure reliable offline data collection, three essential requirements must be addressed: equipment and environmental stability, time synchronization and data consistency, as well as accurate annotation and efficient storage management.

First, appropriate hardware selection and a stable physical environment are necessary for capturing valid and reproducible CSI. Commonly used platforms include Intel 5300 network interface cards and SDR devices equipped with multi-antenna configurations (e.g., 3×3 MIMO), which allow for fine-grained signal acquisition. Proper installation of drivers and firmware ensures signal fidelity and prevents data corruption. Since CSI is highly sensitive to ambient changes such as human activity or electromagnetic interference, data collection should be carried out in controlled indoor environments with minimal disturbance. Maintaining stable parameters like antenna placement, transmission power, and frequency bands (e.g., 2.4, 5, or 6 GHz) helps to ensure consistency across sessions. In addition, calibration procedures should be performed before data collection to compensate for device-related variability. Preprocessing techniques such as filtering, denoising, and amplitude normalization further improve data quality and facilitate downstream analysis.

Second, precise time synchronization is critical, especially in multi-device collection setups where the alignment of CSI measurements must be maintained across all nodes. Asynchronous timestamps may introduce inconsistency and lead to misaligned input-output pairs in supervised learning tasks. Protocols such as NTP and PTP can provide high-precision synchronization, ensuring temporal consistency during acquisition. Moreover, maintaining a stable and sufficiently high sampling rate (e.g., 100 or 200 Hz) ensures dense and uniformly distributed data, which is particularly important for modeling dynamic movements. To address unavoidable issues such as packet loss or jitter during wireless transmission, appropriate strategies including interpolation, frame alignment, and temporal filtering should be applied to recover data continuity and suppress abrupt signal variations.

Third, accurate annotation and robust data management are indispensable for supervised learning-based positioning methods. Ground truth location data should be collected using high-precision tools such as UWB positioning systems, laser rangefinders, or optical motion capture systems (e.g., VICON), which can achieve centimeter-level accuracy. These systems provide reliable references for model training and evaluation. Meanwhile, raw CSI measurements typically require a series of preprocessing steps, including noise reduction, normalization, and format standardization, to ensure compatibility and consistency across datasets. Commonly used storage formats include CSV, MATLAB .mat files, and HDF5, which facilitate efficient data access, sharing, and reuse. To prevent data loss or corruption, systematic backup mechanisms and integrity verification should also be implemented as part of the storage protocol.

3.1.3 How to Balance the Data Quality and Quantity

In intelligent localization systems based on ISAC, offline data collection is a critical step for constructing effective machine learning models. A major challenge in this process lies in managing the trade-off between data quality and data quantity. High-quality data enhances model accuracy and robustness, while sufficient data volume supports generalization across diverse environments. To address this, a practical strategy of limited manual collection combined with automated global prediction can be adopted, as illustrated in Fig. 3.2. This approach begins with a small set of high-quality samples obtained through manual collection, followed by the use of automated methods to expand the dataset through model-driven prediction and acquisition (Zhu et al. 2020).

In the initial phase, manual collection ensures a reliable foundation. Professionals collect a limited amount of high-precision data under controlled conditions, with careful calibration of target positions and environmental parameters. This manually curated dataset is used to train a preliminary model with high accuracy, even though its spatial coverage is limited. For example, manually annotated ground truth can effectively mitigate measurement errors caused by device drift or environmental noise in CSI-based localization. Once the initial model is constructed, automated methods are used to expand the dataset. Automated sensing systems such as mobile robots, sensor networks, or wearable devices can continuously collect data across larger areas. By applying the preliminary model to filter and predict the newly acquired data, the system can retain only meaningful samples that align with quality standards. Additionally, adaptive strategies such as reinforcement learning

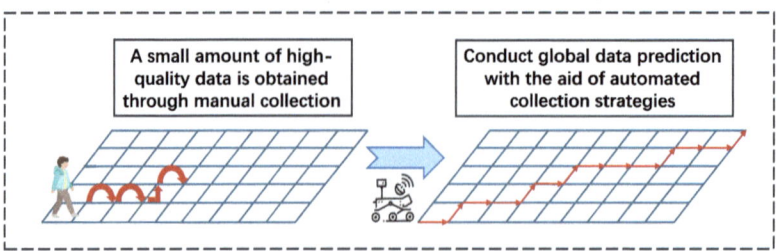

Fig. 3.2 Balancing data quality and quantity through hybrid collection

or entropy-based sampling can guide automated systems to focus on regions with uncertain or underrepresented data, thereby improving coverage while controlling redundancy.

Combining manual and automated methods enables an efficient and scalable collection framework. In complex or highly variable regions such as corners, doorways, or environments with strong multipath effects, manual collection can be prioritized to ensure data accuracy. In contrast, in stable environments like hallways or open areas, automated collection can improve speed and efficiency. Where necessary, manual verification can be introduced to refine key samples obtained through automation, ensuring consistency across the dataset. This hybrid strategy not only preserves data quality but also improves efficiency, supporting the scalable development of intelligent localization systems. In real-world deployments, the collection strategy should be dynamically adjusted based on environmental complexity, system objectives, and resource constraints to ensure optimal dataset construction.

3.2 Automated Data Collection Systems: Reducing Manual Effort

3.2.1 Robotic and Autonomous Data Collection Devices

With the advancement of indoor localization technology, the demand for high-quality data has significantly increased. Traditional manual data collection methods, while effective in some cases, are time-consuming, labor-intensive, and prone to human error. As the complexity of localization systems grows, these conventional approaches are increasingly insufficient to meet the needs of modern applications. The introduction of automated data collection systems, specifically robotic devices, can alleviate human workload and substantially improve both data efficiency and consistency. Robotic devices are especially valuable in complex or hazardous environments where they can replace human efforts, performing tasks that require precision and safety. Common robotic devices such as mobile robots, unmanned aerial vehicles (UAVs), and automated sensor platforms offer versatile and effective solutions for data collection across a wide range of environments and scenarios, ensuring that the data collected is both accurate and comprehensive (Zheng et al. 2025).

Mobile robots, equipped with advanced sensors such as LiDAR, cameras, and wireless signal receivers, are capable of autonomously navigating both indoor and outdoor environments to collect data. These robots can be deployed in a variety of settings, including warehouses, factories, or large office buildings, where they autonomously traverse predefined paths to gather real-time CSI and environmental data. Their ability to maneuver around obstacles and adapt to changes in the environment enables them to maintain consistent data acquisition even in complex or dynamic conditions. For instance, in indoor positioning scenarios, mobile robots can continuously collect data along predetermined paths, ensuring precise and repeatable measurements without the need for manual intervention. UAVs, in contrast, are ideal for large-scale or hard-to-reach areas such as high-rise buildings, forests, or disaster sites, where their ability to navigate through the air allows them to access locations that are otherwise difficult or unsafe for humans to reach. Equipped with high-precision sensors and communication modules, UAVs can rapidly gather data from these environments and transmit it wirelessly to central systems for analysis. Finally, automated sensor platforms, often deployed at fixed locations, can enable continuous data acquisition without requiring human oversight. In settings like smart cities, these platforms can monitor various environmental parameters such as air quality, traffic flow, and wireless signal strength, facilitating real-time data collection over extended periods.

The main advantages of robotic devices lie in their flexibility, programmability, and autonomy. Through the use of intelligent algorithms such as path planning, obstacle avoidance, and task scheduling, robotic devices can dynamically adjust their collection strategies based on real-time environmental conditions, maximizing both efficiency and coverage. For example, if a robot encounters an obstacle, it can autonomously reroute itself to ensure that data collection continues without interruption. This dynamic adaptability allows robots to work in environments where conditions are constantly changing. Furthermore, robotic data collection systems can follow predefined paths or perform specific tasks, allowing for more targeted data acquisition without constant human supervision. Their stable speed and posture further enhance data consistency and repeatability, which is particularly critical in ensuring the reliability of localization models. Additionally, many robotic systems support remote monitoring and control, providing researchers with the ability to adjust experimental parameters in real time to optimize data collection processes. This flexibility enhances the overall efficiency of the data collection, ensuring that large-scale, high-quality datasets can be obtained quickly and reliably with minimal human involvement.

3.2.2 AI-Powered Collection and Sampling

AI technologies have become increasingly valuable for improving the efficiency and adaptability of data collection with the growing complexity of indoor localization systems. Traditional methods often follow fixed sampling rules and predetermined paths, which limits their ability to respond to dynamic environments or changing signal conditions. In contrast, AI-based approaches rely on data-driven strategies to enhance data quality, improve coverage, and reduce manual intervention. This section introduces typical applications of reinforcement learning, deep learning, and swarm intelligence in automated data collection.

3.2 Automated Data Collection Systems: Reducing Manual Effort 53

Intelligent Path Planning and Adaptive Sampling Using Reinforcement Learning Reinforcement learning (RL) improves data collection by optimizing movement paths and sampling frequencies based on real-time environmental feedback. In indoor localization scenarios, mobile robots can use RL to increase sampling density in areas where signals are unstable or multipath effects are prominent. Through repeated interaction with the environment, robots gradually learn efficient paths that ensure better coverage and more useful data. RL also allows systems to adjust sampling rates dynamically, collecting more data where variation is high and less where signals remain stable. This helps reduce redundancy while maintaining high-quality data.

Signal Prediction and Sampling Refinement Using Deep Learning Deep learning (DL) supports efficient data collection by learning signal patterns from historical measurements. Models such as CNN and RNN can forecast signal strength changes and identify important sampling points. For example, DL can help estimate spatial features such as signal attenuation or reflection based on previous CSI data, enabling more targeted and efficient sampling. These models can also detect abnormal patterns and filter out noisy data during collection, contributing to higher data reliability under varying environmental conditions.

Coordinated Multi-Device Collection Using Swarm Intelligence In large-scale data collection tasks, coordination between multiple devices is necessary for efficiency. Swarm intelligence algorithms, including ant colony optimization and particle swarm optimization, offer effective solutions for distributed task scheduling and cooperative path planning. For instance, a group of robots or drones can divide a sensing area into regions and assign tasks based on workload and proximity. These algorithms enable decentralized collaboration, reduce overlap, and improve the overall speed and accuracy of data collection in complex environments. In the future, with the continuous development of AI technologies, data collection systems will become more intelligent and adaptive, capable of addressing more complex scenarios and higher precision requirements.

3.2.3 *Streamlining Collection for Large Datasets*

As dataset size increases, traditional collection methods often struggle with low efficiency, high labor costs, and growing complexity. To address these challenges, robotic platforms and AI-driven strategies can be jointly applied to optimize the process. One effective approach is to combine optimal path planning with data prediction. Robotic devices such as mobile robots or drones collect data along intelligently generated paths, ensuring dense sampling in complex regions while reducing redundancy in stable areas. During operation, these devices can dynamically adjust sampling strategies to prioritize high-value data. Based on collected samples, models such as Gaussian process regression or neural networks are used to predict unsampled regions, effectively expanding the dataset while lowering collection effort and post-processing costs.

Scaling up data collection requires coordination across multiple robotic devices. These systems can be flexibly deployed to cover wide areas, with each device autonomously adjusting its strategy based on real-time feedback or task priorities. In smart city scenarios, for instance, drones can divide regions for parallel collection, while ground robots continuously operate through automated charging and scheduling systems. Swarm intelligence algorithms such as ant colony or particle swarm optimization further enhance task allocation and route coordination, improving coverage efficiency and avoiding resource conflicts. This enables large-scale data acquisition with reduced human involvement and higher overall consistency.

AI techniques such as reinforcement learning and deep learning are essential for intelligent sampling and system optimization. Reinforcement learning adjusts sampling behavior based on environmental feedback, increasing frequency in dynamic areas while reducing redundancy in stable regions. Deep learning models, trained on historical CSI data, predict signal trends and guide path selection, enabling more targeted and efficient sampling. At the same time, distributed data management systems like Hadoop or Spark facilitate real-time processing across sensor networks. Preprocessing and compression techniques, such as denoising, normalization, and lightweight encoding, help minimize storage requirements and transmission overhead. Together, these technologies optimize large-scale data collection, ensuring quality, scalability, and adaptability in complex environments.

3.3 Device-Based vs. Device-Free Collection Approaches

3.3.1 Device-Based Collection: Pros and Cons

Data acquisition methods for indoor positioning are generally categorized into device-based and device-free collection approaches. Device-based collection requires targets, such as users or objects, to carry specific sensing devices, including smartphones, wearable devices, or dedicated sensors. These methods have several advantages but also certain limitations that make them more suitable for specific applications.

Common device-based methods include smartphone sensing, wearable devices, and UWB tags (Zhou et al. 2024). Smartphone sensing, which utilizes built-in sensors like Wi-Fi, Bluetooth, and IMU, is one of the most widely used techniques, especially for indoor navigation and activity recognition. Wearable devices, such as smartwatches and smart glasses, offer additional functionality by providing data on motion, health, and user activity. These are valuable in applications like health monitoring and activity tracking. UWB technology, known for its high-precision positioning capabilities, is frequently deployed in settings that demand stringent accuracy, such as industrial environments and medical facilities.

3.3 Device-Based vs. Device-Free Collection Approaches 55

A notable application of device-based collection is crowdsourcing, where a large number of users contribute data via their personal devices, collected at various times and locations. The data is then uploaded to the cloud for aggregation and analysis. The key advantage of this approach is its ability to collect large volumes of data over extensive areas at relatively low cost. This makes crowdsourcing particularly effective for indoor positioning and sensing in large-scale environments. However, the data quality can vary significantly due to factors like device heterogeneity and inconsistencies in user behavior. Furthermore, privacy and security issues arise, as sensitive user information may be collected, requiring proper anonymization and encryption techniques to ensure protection.

While device-based collection methods offer high-quality data with direct links to targets, they also come with limitations. These methods depend on users carrying devices, which can lead to gaps in data collection, as participation may not be continuous or universal. The limited number of devices available also restricts the coverage of data. Additionally, continuously running sensors consume considerable energy, raising concerns about battery life and device maintenance. Moreover, the collection and transmission of personal data introduce privacy risks, necessitating robust security measures, including encryption and anonymization. Therefore, although device-based methods provide precise data, there are challenges related to user participation, data coverage, energy consumption, and privacy that must be addressed for broader applicability.

3.3.2 *Device-Free Sensing for Localization*

Device-free localization refers to positioning techniques that do not require the target to carry any specific device. This approach improves convenience, especially in environments where user cooperation is limited or infeasible. By leveraging environmental sensors or signal variations, device-free methods can achieve accurate human localization in a wide range of indoor scenarios. Among the various techniques, spatial video-based human localization has gained increasing attention due to its high precision and intuitive spatial awareness.

Spatial video-based localization uses visual sensing systems such as RGB-D cameras, stereo vision, or multi-camera arrays to capture three-dimensional information about the environment. These systems analyze human posture, movement, and spatial occupancy in real time to determine the position of individuals. Unlike traditional video surveillance, spatial video systems are designed to extract geometric and motion features for precise localization. By integrating depth information and advanced computer vision algorithms, such as pose estimation and semantic segmentation, these systems can localize people even in complex indoor settings with occlusions or dynamic scenes. This method is particularly suitable for scenarios that require high precision and non-intrusiveness, such as elderly monitoring, security surveillance, and human-robot interaction.

In addition to spatial video, a range of other device-free localization methods have been developed based on wireless signals. Techniques using Wi-Fi, UWB, and Bluetooth rely on changes in signal strength, CSI, or time-of-flight measurements to infer human presence and location. For example, Wi-Fi-based sensing can detect human motion by analyzing the disturbances caused by the human body in wireless signal propagation. UWB systems offer higher resolution by measuring signal travel time with great accuracy, making them suitable for industrial or healthcare environments. However, these methods can be sensitive to multipath effects and environmental interference, which may limit their consistency in complex scenarios. Acoustic sensing also represents a class of device-free techniques, using ultrasonic or audible signals for localization. By analyzing the reflection and delay of sound waves, these systems can estimate the position of people in confined spaces. While offering high precision under controlled conditions, acoustic methods often face challenges such as background noise and the need for LoS signal paths.

3.3.3 Minimizing Data Loss During Collection

In offline data collection, data loss can be caused by various factors, such as signal attenuation, device failures, network instability, or human intervention. To enhance data integrity, preventive methods can be implemented before data acquisition, and data processing techniques can be applied afterward to compensate for and recover lost data.

Before data collection, the possibility of data loss can be minimized through proper system design and hardware optimization. Key methods include: dynamic sampling rate adjustment, as different environments have different requirements for sampling rates. An excessively high sampling rate may lead to redundant data, while a low sampling rate may result in information loss. Dynamically adjusting the sampling rate based on signal quality can optimize data storage and transmission efficiency. Real-time data integrity checks should also be conducted during data collection, where devices can monitor data integrity in real-time, such as tracking the packet loss rate of Wi-Fi or detecting abnormalities like sudden changes or prolonged inactivity in IMU data. Additionally, optimal sensor deployment should be implemented. For device-free data collection scenarios, such as with Wi-Fi access points, RFID readers, or millimeter-wave radars, optimizing their layout can reduce signal blind spots and enhance signal coverage and stability.

Even with preventive methods, data loss or anomalies may still occur during collection, necessitating post-collection data repair. Common methods for addressing this issue include interpolation algorithms for data repair, such as linear interpolation, spline interpolation, or Kalman filtering, which are used to fill in short-term gaps in time-series data (e.g., IMU sensor data or Wi-Fi). For data with low signal quality, denoising and enhancement techniques like wavelet transform, PCA, or denoising autoencoders can help reduce noise and improve the usability of the data (Liu et al. 2025). Additionally, data fusion and compensation can be applied when data from one sensor is lost; for instance, if Wi-Fi signal quality is poor, IMU data can be used to infer target motion trajectories, or Bluetooth RSSI data can assist in positioning.

In summary, minimizing data loss requires a two-pronged approach: optimizing pre-collection processes and implementing post-collection compensation. Before data acquisition, stability can be improved by dynamically adjusting sampling rates, monitoring data integrity in real-time, and optimizing sensor deployment. After collection, missing data can be repaired using interpolation, signals can be enhanced through noise reduction, and multi-sensor information can be fused to mitigate the impact of data loss on positioning accuracy. By optimizing each stage, data integrity can be ensured, thereby enhancing the reliability of subsequent analysis and positioning. Additionally, manual, semi-automated, and fully automated CSI data collection methods differ in terms of human effort, scalability, cost, and applicable scenarios, and should be selected according to requirements, resources, and budget, as listed in the Table 3.1.

3.4 A3C-IPP Algorithm

3.4.1 Preliminaries

Fingerprinting is a common indoor positioning technology, and its offline phase requires the construction of a fingerprint database. This involves dividing the experimental area into grids and measuring data such as CSI at each sampling point to form a comprehensive fingerprint map. However, this process relies on manual data collection, which is time-consuming and labor-intensive, especially in large-scale environments where the workload increases significantly. Therefore, how to reduce labor costs while ensuring the accuracy of the fingerprint database has become a critical challenge. In recent years, the application of robotics technology has provided an automated solution, where intelligent agents equipped with mobile sensing devices can autonomously collect data and possess obstacle avoidance capabilities. However, due to limited battery life, it is necessary to optimize their path planning to ensure efficient data collection within constrained resources.

The path planning problem is typically modeled as the Traveling Salesman Problem (TSP). Traditional methods such as genetic algorithms and simulated annealing, while capable of providing solutions, are prone to falling into local optima and suffer from high computational complexity. Reinforcement learning approaches, such as DDPG, DQN, and Q-learning, can be applied to path optimization but face challenges such as sparse rewards and low training efficiency. Furthermore, although UAV offer high mobility, factors such as flight instability and signal fluctuations in complex indoor environments can degrade the quality of CSI collection. To address these challenges, this section proposes the A3C-IPP (Asynchronous Advantage Actor-Critic-based Indoor Path Planning for fingerprint

Table 3.1 Comparison of methods for reducing manual effort in CSI data collection

Comparison criteria	Manual collection	Semi-automated collection	Fully automated collection
Human Involvement	Full-time human operation at each sampling point; requires manual annotation and path tracking	Human assistance during robot navigation, calibration, and troubleshooting; reduced manual data labeling	Minimal involvement after initial deployment; only required for maintenance and occasional reconfiguration
Data Collection Speed	Low; limited by walking speed and manual recording; inefficient for large areas	Medium; robots follow pre-defined routes at steady speed; human guidance may still be needed in complex environments	High; autonomous agents collect data in parallel and adapt to environmental changes without human delay
Accuracy and Consistency	Varies due to human inconsistency and potential positioning errors; hard to reproduce exact paths	Better reproducibility due to programmed paths and sensor fusion; moderate deviation in dynamic scenes	High spatial and temporal consistency; adaptive algorithms adjust routes and sample density in real time
Equipment Requirements	Handheld devices such as laptops, smartphones, or Wi-Fi sniffers; simple tools but labor-intensive	Robots or motorized trolleys with mounted sensors (e.g., Wi-Fi modules, IMUs); moderate hardware setup	Drones, mobile robots, or fixed sensor arrays with advanced onboard processing and wireless modules
Calibration and Quality Control	Manual calibration before/after collection; results checked post-hoc; time-consuming and error-prone	Robot-assisted calibration with limited real-time feedback; periodic human validation still needed	Self-calibration using sensor feedback; online anomaly detection ensures real-time quality control and reliability
Scalability	Practical only for small labs or single-room settings; significant human labor for expansion	Feasible for building-scale deployments; performance depends on number of available robots and coverage plan	Easily extensible to campus or city-scale with proper infrastructure; supports large-scale parallel operations
Energy and Resource Consumption	High energy demand from personnel; collection is time-intensive and often repeated	Moderate energy use; robots can operate longer with fewer breaks; still needs occasional human recharge intervention	Energy-efficient path planning and scheduling; autonomous recharging and task handover reduce human workload
Cost Structure	Low initial equipment cost but high cumulative labor expenses; inefficient for recurring tasks	Moderate hardware investment balanced by reduced manpower; requires occasional technical support	High initial deployment and system integration cost; operational cost is low due to automation and reusability

3.4 A3C-IPP Algorithm

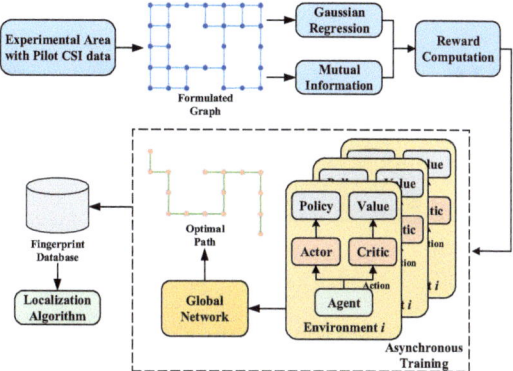

Fig. 3.3 The framework of the A3C-IPP algorithm

map construction) algorithm. By utilizing the asynchronous multi-agent learning capabilities of A3C, the algorithm enhances path search efficiency and optimizes CSI collection paths in dynamic environments, thereby improving the efficiency and accuracy of fingerprint database construction.

3.4.2 The Overview

The overall framework of the A3C-IPP algorithm is shown in the Fig. 3.3, which consists of three core modules: reward value calculation based on multivariate Gaussian regression, path exploration strategy based on A3C, and online localization. Initially, a small number of sampling points conforming to a normal distribution are randomly selected within the target area to collect initial CSI fingerprint data. Subsequently, the divided grid is modeled as a graph structure, and the physical coordinates of the sampling points along with the initial data are used to estimate the CSI data distribution via a multivariate Gaussian regression model. The reward value is then calculated based on mutual information.

Next, the path planning problem is transformed into a linear decision-making problem, and the A3C algorithm is employed to search for the optimal path. A3C enhances training efficiency through multi-threaded parallel computing and utilizes a CNN to share global model parameters, thereby improving stability. Additionally, a novel reward mechanism is introduced to more effectively guide the agent in exploring the optimal path.

Finally, only the CSI fingerprint data along the optimal path is collected, and the CSI distribution of unsampled points is corrected based on this data to construct a complete CSI fingerprint database, which supports subsequent localization tasks. CSI provides richer physical layer information compared to RSSI, enabling higher-precision localization. Existing research has verified that CSI-based indoor localization methods can achieve cm-level accuracy, whereas RSSI-based methods exhibit localization errors of approximately 2.7 m. Therefore, using CSI for path

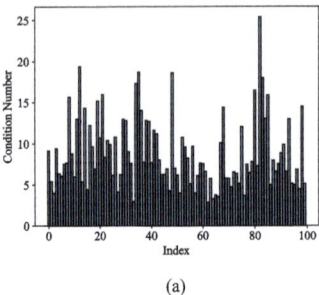 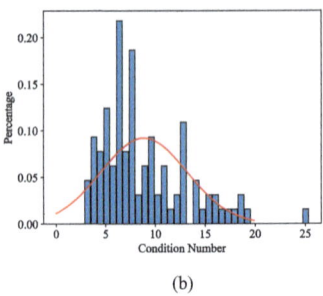

Fig. 3.4 CSI collinearity analysis. (**a**) Condition number. (**b**) Distribution statistics

planning is expected to further enhance localization accuracy. The core objective of this algorithm is to reduce the cost of constructing the fingerprint database during the offline phase, while the online localization phase employs the KNN algorithm for position estimation to validate the effectiveness of the fingerprint database.

The proposed algorithm constructs a CSI fingerprint database based on the path loss model, assuming a linear relationship between fingerprint measurements. However, real-world factors like reflection and attenuation may cause multi-collinearity. The condition number is used to assess this: values above 10 indicate multicollinearity, and above 30 indicate severity. As shown in Fig. 3.4, CSI data exhibit low collinearity, supporting the use of multivariate linear regression for distribution prediction.

3.4.3 Reward Computation

The target area is discretized into a grid graph, where vertices represent the physical coordinates of sampling points, and edges denote the line segments between directly connectable points, with edge rewards calculated via an approximation algorithm. Due to multipath effects, signal attenuation, and environmental changes, directly modeling fingerprint signals proves challenging. Therefore, this algorithm employs a multivariate Gaussian process to model the relatively stable CSI fingerprint signals. For $\{f(x) : x \in \mathcal{X}\}$, the model can be established as follows: $f(\cdot) \sim GP(m(\cdot), k(\cdot))$.

Let $m(\cdot)$ be the mean function and $k(\cdot)$ be the covariance function. Thus, for all $x, x' \in \mathcal{X}$, we have $m(x) = \mathbb{E}[x]$ and $k(x, x') = \mathbb{E}[(x - m(x))(x' - m(x'))]$. Regarding the path planning problem, we denote the target location as $\mathcal{X}$, meaning that the distribution of CSI data depends on the actual physical coordinates.

3.4 A3C-IPP Algorithm

Fig. 3.5 Mutual information relationship

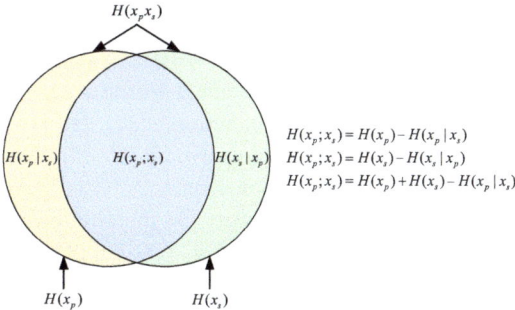

Specifically, the directional basis function is chosen as the kernel function for computation, and the formula is given as

$$k(x, x') = \exp\left(-\frac{1}{2\tau^2} \|x - x'\|^2\right). \tag{3.1}$$

Using a small set of pre-collected real CSI data (x_p), we compute the reward values for all edges based on mutual information. Given the full dataset (x_s), the mutual information is defined by the conditional distribution $p(x_p|x_s)$, where a larger deviation between $p(x_p|x_s)$ and the mean of x_p indicates higher information gain. The conditional distribution is transformed into differential entropy H, and the mutual information computation is shown in Fig. 3.5. This approach enables estimating the overall data distribution with minimal initial data, and our method employs the first equation to compute $H(x_p; x_s)$.

The differential entropy of a Gaussian distribution is calculated as

$$H(x) = -\int f(x)\log\phi(x)\,dx = \frac{n}{2}\log\left(2\pi e\sigma^2\right), \tag{3.2}$$

where σ depends on the covariance function $k(\cdot)$, while all other parameters are constants. If the CSI measurements of all sampling points are denoted as y_s, then $H(x_p)$ can be directly computed from the above equation, and the covariance matrix Σ can be expressed as

$$\Sigma = k(x_p, x_p) - k(x_p, x_s)\left(k(x_s, x_s) + \hat{\sigma}_n^2 I\right)^{-1} k(x_s, x_p), \tag{3.3}$$

where $\hat{\sigma}_n$ represents the noise. Then, the $H(x_p|x_s)$ can be computed as

$$H(x_p \mid x_s) = \frac{n}{2}(\ln 2\pi + 1) + n\ln\hat{\sigma}. \tag{3.4}$$

Finally, the reward value R based on mutual information can be computed as $R = H(x_p; x_s) = H(x_p) - H(x_p|x_s)$.

3.4.4 Exploration Strategy Based A3C

The framework of reinforcement learning algorithms consists of five key components: agent, environment, action, state, and reward. This study employs the A3C algorithm for path planning, which is well-suited for handling high-dimensional data and continuous action spaces. A3C utilizes multiple asynchronous agents to explore different state transitions, thereby reducing correlations between samples during training. Moreover, A3C can be executed on multi-core CPUs, providing advantages over traditional methods in terms of performance, time efficiency, and resource consumption.

A3C follows an Actor-Critic framework, where the Actor optimizes the policy $\pi(a_t|s_t;\theta)$ to improve decision-making, while the Critic updates the value function $V(s_t;\theta_v)$ to enhance accuracy. The policy update method is given as

$$\nabla_{\theta'} \log \pi \left(a_t \mid s_t; \theta'\right) A(s_t, a_t; \theta, \theta_v), \tag{3.5}$$

where $A(s_t, a_t; \theta, \theta_v)$ is the advantage function, which can be expressed as

$$A(s_t, a_t; \theta, \theta_v) = \sum_{i=0}^{k-1} \gamma^i r_{t+i} + \gamma^k V(s_{t+k}; \theta_v) - V(s_t; \theta_v), \tag{3.6}$$

where θ is the parameters of the policy π, and θ_v is the parameters of the value function.

The following sections provide a detailed implementation process for path planning using A3C. First, according to the requirements of path planning, we give the specific definitions as follows:

- **Agent**: The agent, modeled as a robot, explores an optimal path from the start S to the goal G. After each training iteration, it resets to S and attempts to reach G within a fixed step limit.
- **Environment**: Represented as a graph of vertices and edges based on spatial layout, the environment includes LoS and NLoS scenarios for localization.
- **Action**: The agent explores different actions, moving along the edges between vertices. The final strategy consists of a sequence of selected actions.
- **State**: The state represents the agent's position and action sequence, requiring obstacle avoidance. A3C uses a global network with shared parameters to optimize action selection.

(continued)

3.4 A3C-IPP Algorithm

> - **Reward**: After executing an action, the agent receives a reward value, which can be positive or negative. As the agent moves progressively closer to the goal G, the cumulative global reward increases accordingly.

RL-based path planning faces two major challenges: obstacle occlusion and algorithmic efficiency. Firstly, obstacles may prevent direct connections between adjacent points, necessitating obstacle avoidance solutions. Secondly, the high computational complexity of the algorithm requires limiting iteration steps to ensure efficiency. The optimal path is defined as the one with the maximum reward value and information gain, rather than the shortest distance.

(1) Exploration Strategy The start point S and the goal point G are set at diagonal positions. In theory, data collected along the diagonal following a Gaussian distribution can predict the distribution of other points. The action space includes "up," "down," "left," and "right," but a complex environment may render the diagonal ineffective. Suppose the maximum exploration step length max_step is much smaller than the total number of sampling points. If the agent reaches G within max_step, the path is considered found; otherwise, the agent is reset to S to reinitiate the exploration.

Let all potential paths be denoted as $P = [v_S, \ldots, v_G]$, with the corresponding total reward value denoted as $r(P)$. The search for the optimal path should satisfy

$$P_{\text{optimal}} = \arg\max_{P \in \Psi} r(P), \tag{3.7}$$

where Ψ represents the set of all potential paths from v_S to v_G. If v_i denotes the current position of the agent, then the available action set is given by

$$A(v_i) = \{v_{i+1} \in V : (v_i, v_{i+1}) \in E\}, \tag{3.8}$$

where V and E denote the sets of vertices and edges, respectively. The action set A depends on the adjacent position v_{i+1}. In traditional methods, the agent randomly selects an action from A. If the resulting state is an obstacle, the agent is reset for a new iteration.

The proposed algorithm argues that the method wastes action selection cycles and involves a computationally expensive execution process. The A3C-IPP algorithm introduces a rollback mechanism and a greedy strategy for action selection. Here, the proposed algorithm adopts a greedy strategy similar to the Q-learning algorithm to select the optimal action. The reward value obtained in each iteration is computed as follows. As the iterations progress, the total reward value is updated by $r(v_{i+1}) = r(v_i + A(v_i)) - r(v_i)$.

(2) Optimal Path To avoid local optima and accelerate convergence, the A3C algorithm employs multithreading to search for the optimal policy. Another optimization approach is to introduce an entropy term with a coefficient β into the actor-critic policy loss function, as defined by Eq. (3.5), which can be expressed as $\theta = \theta + \alpha \nabla_\theta \log \pi_\theta(s_t, a_t) A + \beta \nabla_\theta H(\pi(s_t, \theta))$.

The global network model of A3C consists of both the Actor and Critic neural networks. It utilizes n worker threads, each with the same structure as the global network, and independently interacts with the environment to collect experience data. After each interaction, each thread computes the local gradient, where the update gradient for the Critic network is given by

$$d\theta \leftarrow d\theta + \frac{\partial (r - Q(s, a; \theta'))^2}{\partial \theta'}. \tag{3.9}$$

Particularly, these gradients are used to update the global network rather than the thread-specific networks. Each of the n threads independently accumulates gradients and updates the parameters of the global network, while periodically synchronizing its own parameters to guide subsequent interactions.

By incorporating an exploration strategy, the agent can identify multiple potentially effective paths. Although A3C gradually increases the accumulated reward through the global neural network, it does not guarantee finding the optimal solution. Therefore, the proposed algorithm adopts a greedy strategy, computing the reward value for each path and selecting the path with the highest total reward as the final solution based on Eq. (3.7).

3.4.5 Predict the Distribution of Fingerprints

By collecting fingerprint data along the optimal path and integrating a small amount of initial raw CSI data, a fingerprint dependency model is constructed to capture the spatial correlation between adjacent locations. Consequently, a global fingerprint database can be established. This algorithm assumes that the spatiotemporal distribution of fingerprint data depends on the actual physical coordinates. The real-time collected optimal path fingerprints are used as model inputs to predict the fingerprint data distribution at other locations. Specifically, the invariance of the dependency model serves as a fundamental prior assumption for constructing the entire fingerprint map.

The covariance function $k(\cdot)$ is the core module for predicting the fingerprint distribution. Given n sampled locations (including those used for the initial Gaussian model construction and the key vertices along the optimal path) with corresponding CSI data as

$$CSI_j = \sum_{j=1}^{m} \sum_{i=1}^{n} \text{cov}(i, j) \times CSI_i, \tag{3.10}$$

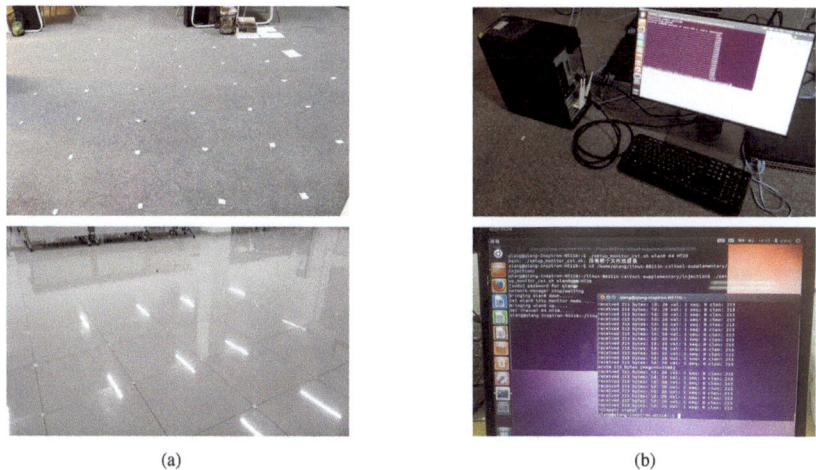

Fig. 3.6 The photos of scenarios and devices. (**a**) The scenarios. (**b**) The devices

where cov(·) represents the covariance matrix. The physical coordinates (i.e., labels) of all the sampling points in the localization area are known, and the CSI_i and CSI_j are merged to form the final fingerprint database.

3.4.6 Performance Evaluation

The algorithm employs two devices for experimentation: a Dell XPS desktop (data transmitter) and a Dell laptop (data receiver) as shown in Fig. 3.6. Both devices have the 64-bit Ubuntu 12.04 LTS operating system and Intel 5300 network cards. The network card driver kernel was modified using the Linux 802.11n CSI Tool to capture raw CSI data (Halperin et al. 2010). To validate the algorithm's performance, data collection and experiments were conducted in two representative indoor environments: a laboratory (Area one), and a meeting room (Area two), as shown in Fig. 3.7.

For **Area one**, a $13.5 \times 11\,m^2$ computer laboratory cluttered with tables, chairs, and obstacles, representing a **NLoS** scenario where movement introduces additional uncertainty, thus demanding algorithm robustness, featuring 317 sampling locations spaced 50 cm apart with the transmitter and receiver placed on the floor for continuous packet reception. For **Area two**, a $7 \times 10\,m^2$ nearly empty meeting room with minimal signal shielding and no human movement during data collection, serving as a **LoS** scenario with 176 sampling locations spaced 60 cm apart, where pilot data were gathered for several seconds at each location, using 1000 continuous packets to ensure data validity.

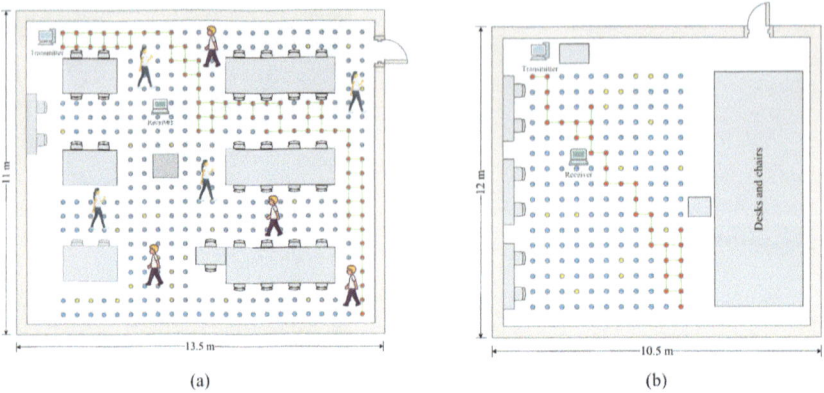

Fig. 3.7 The experimental scenarios. (**a**) Layout of area one. (**b**) Layout of area two

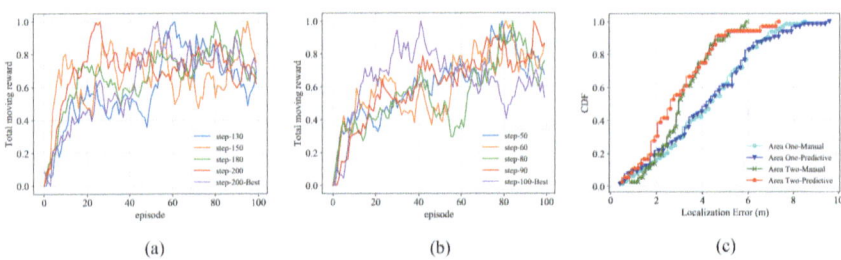

Fig. 3.8 Total moving reward with A3C and localization accuracy. (**a**) Area one. (**b**) Area two. (**c**) CDF of localization error

(1) Training and Localization Using Optimal Path First, we evaluate the convergence performance with different maximum exploration step lengths, keeping other parameters fixed. Since larger step sizes yield higher total rewards, we normalize the reward values for fair comparison. Figure 3.8a, b show the moving rewards in both scenarios, over 100 episodes. The A3C-based exploration gradually converges, although early fluctuations appear when the agent fails to reach the goal G. Compared to Area two, Area one shows more reward oscillations due to greater environmental complexity and dynamic interference, occasionally causing the agent to fall into local optima. Start and goal vertices are placed diagonally, and with increased step size, the agent explores more positions. The path achieving the highest reward under the maximum exploration length is selected as the final output.

Next, we compare localization accuracy between the predictive fingerprint database and manually collected data. Full CSI datasets were gathered in both areas, and KNN is used for localization to focus on database effectiveness. At each location, over 3000 packets were collected, and 1000 were used for training with cross-validation on randomly split datasets. Figure 3.8c shows that both databases achieve similar localization performance. In Area one, the predictive database

3.4 A3C-IPP Algorithm

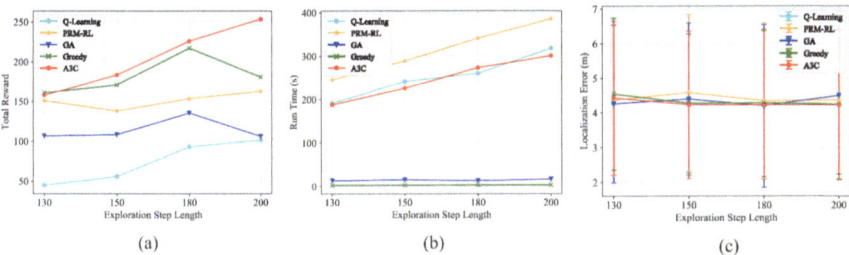

Fig. 3.9 The performance comparison in area one. (**a**) Total reward. (**b**) Run time. (**c**) Localization error

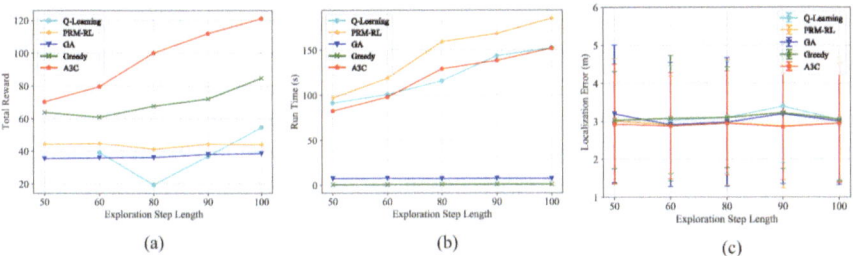

Fig. 3.10 The performance comparison in area two. (**a**) Total reward. (**b**) Run time. (**c**) Localization error

achieves 9% of errors within 1.0 m compared to 3% for the manual data, and mean errors are 4.22 and 4.22 m, respectively. In Area two, mean errors are 2.95 m for the predictive and 3.20 m for the manual database. Importantly, our method reduces data collection by over 68% while maintaining comparable accuracy. Further improvements are possible with techniques like Kalman filtering and principal component analysis.

(2) Comparison with Other Algorithms We compare the proposed A3C-IPP with the several existing methods, including Q-Learning (Low et al. 2019), PRM-RL (Faust et al. 2018), GA (Wei et al. 2019), and Greedy (Wei et al. 2019). As shown in Figs. 3.9 and 3.10, the proposed algorithm consistently demonstrates superior performance across different exploration step lengths. Compared with Q-learning, which struggles with early-stage action selection due to inaccurate Q-values, the proposed method steadily improves total rewards and localization accuracy. Q-learning often converges to local optima and exhibits higher localization errors. PRM-RL, based on the actor-critic framework, succeeds in identifying feasible paths but requires longer training time and shows slower convergence, despite maintaining relatively stable localization performance. The Genetic Algorithm, which evolves valid paths generated by A3C, achieves faster computation but suffers from unstable localization results and limited reward growth due to frequent local optima. The Greedy Algorithm, by prioritizing immediate rewards, achieves high efficiency but cannot guarantee reaching the target, leading to inconsistent total rewards and lower localization accuracy compared to the proposed approach.

The proposed method, based on A3C and enhanced with a novel exploration strategy, efficiently explores the environment and optimizes the path planning process. In both areas, it shows a steady increase in total rewards and a gradual reduction in localization error with low variance. Although computational time increases in more complex environments, it remains manageable and benefits significantly from GPU acceleration. The positive correlation between total reward and localization accuracy further verifies the effectiveness of the approach.

To summarize, in this chapter, we design an innovative CSI data acquisition method for indoor localization, which leverages a path planning approach within the A3C framework. Recognizing that CSI distributions are inherently position-dependent and approximate a multivariate Gaussian form, we introduce a reward estimation mechanism based on mutual information and differential entropy to guide sampling decisions during the offline phase. Through the asynchronous actor-critic structure of A3C, multiple agents efficiently collaborate to explore optimal paths using our tailored action selection strategy. Experimental comparisons confirm that the proposed approach outperforms existing state-of-the-art methods in both efficiency and accuracy. Notably, our method reduces data collection effort by approximately 72% while maintaining localization accuracy on par with manually collected datasets.

3.5 CPPU Algorithm

3.5.1 Motivation and Challenge

Fingerprint-based localization has become a cornerstone technique in ISAC. By utilizing characteristics of wireless signals such as Wi-Fi, fingerprinting enables high-precision indoor localization without requiring additional infrastructure. Despite its advantages, a major limitation lies in the offline phase of constructing the fingerprint database. This process requires extensive manual measurements across dense spatial grids, making it highly labor-intensive and time-consuming. Moreover, the sensitivity of signal features to environmental changes adds further complexity, posing a significant challenge to the scalable deployment of ISAC localization systems.

In response to these challenges, researchers have explored automated strategies aimed at reducing the workload associated with data collection. A commonly adopted approach is robot-assisted path planning, where agents follow predefined trajectories to gather CSI data. Within this framework, Multivariate Gaussian Regression (MGR) or heuristic algorithms are often used to estimate CSI values at unsampled locations, with the goal of minimizing physical measurements. However, Gaussian models typically struggle to capture the nonlinear and spatially diverse

3.5 CPPU Algorithm

patterns present in real-world CSI distributions, leading to reduced prediction accuracy. Heuristic methods also face limitations in adaptability and often yield suboptimal coverage in irregular or dynamic environments.

To overcome the above issues, we propose an algorithm named CPPU, which combines CSI-based path planning with updating. The method first partitions the target region into grids and subregions, followed by collaborative exploration using multiple agents guided by a Spanning Tree Coverage (STC) strategy. This ensures comprehensive coverage with efficient agent coordination. To optimize the acquisition process, dynamic programming is applied to generate compact and efficient data collection paths. In parallel, a generative model based on GAN is used to iteratively learn from collected CSI and predict values at unsampled points, gradually refining the fingerprint database with minimal human effort.

3.5.2 The Overview

The proposed algorithm consists of two phases, informative path planning and CSI data updating, as shown in Fig. 3.11. In the informative path planning phase, the target area is divided into a graph structure composed of vertices and edges, further partitioning into multiple sub-regions. The STC strategy is employed to generate full-coverage paths. Using a reinforcement learning approach, a small number of global sampling points are randomly selected to collect pilot CSI data, and a MGR model is fitted to calculate reward values. By integrating dynamic programming strategies, optimal CSI acquisition paths are generated, thereby reducing the workload of data collection.

In the CSI data updating phase, the CSI data is refined through iterative optimization using a GAN model, enhancing the granularity of the global CSI database. Initially, a coarse-grained CSI database is constructed by integrating real CSI data obtained from optimal paths with the MGR model. The GAN model, comprising a generator and a discriminator, is trained adversarially to generate virtual CSI samples that closely resemble the distribution of real samples. The trained generator model is then utilized to input noise and generate virtual samples, which are compared with actual collected CSI data. Supervised learning techniques

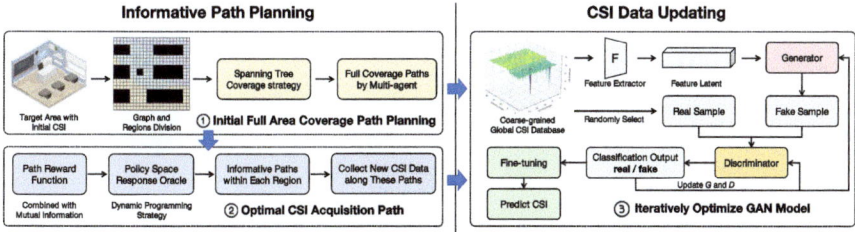

Fig. 3.11 The solution overview of the proposed CPPU

are applied to fine-tune the generator, ensuring the fidelity of the virtual samples. Through multiple iterations, the generator gradually learns to predict and update CSI values at remaining locations with high precision.

The proposed algorithm achieves efficient and accurate indoor localization through informative path planning and CSI data updating. By leveraging multi-agent full-coverage path planning and iterative optimization of the GAN model, the workload of data collection is significantly reduced, while the completeness and accuracy of the CSI database are enhanced.

3.5.3 Problem Formulation

The goal of intelligent path planning is to determine a set of optimal paths in a given environment to collect information-rich data. Formally, the environment can be represented as a graph $G = (V, E)$, where V is the set of vertices representing locations (n sampling points) in the environment, and E is the set of edges representing connections between locations. Each vertex $v_i \in V$ corresponds to a specific location, and each edge $e_{i,j} \in E$ represents the connection between locations v_i and v_j.

In this work, we enhance this problem by introducing three new parameters and further formalizing its description. The new graph is represented as $G = (V, E, Por, Pos, R)$, where:

- $Por = \{p_a, p_b, p_c, \ldots\}$ denotes the sub-regions assigned to different agents, with the sum of their proportions equal to 1.
- Pos represents the initial positions of each agent, expressed as a list of two-dimensional coordinates.
- R denotes the reward values associated with each edge in the graph, which are dynamically updated as agents explore the environment and interact with new CSI data.

The objective of IP is to find a set of paths $P = \{p_1, p_2, \ldots, p_k\}$, where each path p_i is a sequence of vertices $v_1, v_2, \ldots, v_m$, satisfying the following conditions:

- Ensure complete coverage of the environment, with each location visited at least once.
- Enhance the informativeness of the data collected along the paths, providing maximum information about the environment.

(continued)

3.5 CPPU Algorithm

- Avoid revisiting the same location within a single path, ensuring backtrack-free paths.
- Maximize the cumulative reward value obtained from collecting CSI data along the edges, reflecting the value of the collected information.

Driven by the diverse Policy Space Response Oracle (PSRO), the overall optimization objective is defined as

$$\theta_{t+1} = \arg\max_{\theta} \frac{1}{N} \sum_{i=1}^{N} \mathbb{E}_{\sigma_i} \left[U(\theta, \sigma_i) \right], \tag{3.11}$$

where $U(\theta, \sigma_i) = \mathbb{E}_{\sigma_{-i} \sim \pi_{\theta}^{[N-1]}} [U(\sigma_i, \sigma_{-i}) + R(\sigma_i)]$, and $\sigma_i \in \Sigma_i(Por_i, Pos_i, P)$. Here, N denotes the number of agents, σ_i is the strategy of agent i, π_{θ} is the strategy distribution, and U represents the utility function incorporating both interactions and individual rewards.

For the proposed CPPU, we use the Nash Equilibrium (NE) strategy, enabling multiple agents to engage in a game within specific regions to determine the optimal paths that maximize the overall CSI information. Subsequently, a GAN is trained to predict the CSI data at remaining locations. The calculation of the reward value R will be discussed in detail in subsequent chapter. We aim to maximize the similarity between the generated CSI and the real distribution, thereby achieving high-precision CSI updates.

3.5.4 Initial Full Area Coverage Path Planning

The STC algorithm, originally designed for single-agent coverage path planning, offers an efficient method for ensuring optimal coverage paths within a known terrain. Here, we adapt and integrate the STC algorithm to address the challenges outlined in our problem statement. The algorithm is grounded in the concept of Minimum Spanning Trees (MST), where a tree-like structure covering the entire area is constructed to design the agent's path, ensuring each region is covered without redundancy.

Specifically, the terrain is discretized into a finite number of cells, with the total number of cells equal to the grid size. These cells represent the spatial environment in which the coverage path planning algorithm operates. Let the agent's path be denoted as $P_i^* \ \forall i \in \{1, \ldots, N\}$, we get the goal function as

$$\min_{P} \max_{i \in \{1, \ldots, N\}} |P_i| \text{ subject to } P_1 \cup P_2 \cup \cdots \cup P_N \supseteq \mathcal{L},$$

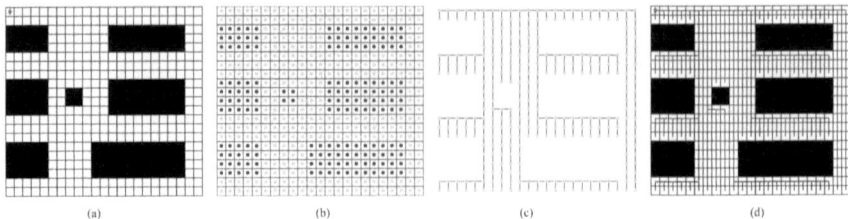

Fig. 3.12 The schematic plot of coverage path planning. (**a**) Cells of terrain, agent, and obstacles. (**b**) Represent the cells as nodes. (**c**) Generate the MST for all the unblocked nodes. (**d**) Utilize the ST to navigate the agent around the terrain

where $|P_i|$ denotes the path length, and $\mathscr{L}$ represents the set of all cells. The STC algorithm generates closed paths by constructing an MST, ensuring coverage of each region.

Figure 3.12 shows the main steps of the trajectory design process. First, the terrain is discretized into cells and abstracted into a graph that distinguishes between traversable and obstacle regions (a). The environment is then grouped into larger cells, each either entirely blocked or navigable, forming the nodes of the graph (b). Edges are added between adjacent unblocked nodes. A MST is then constructed using algorithms such as Kruskal or Prim algorithm (c). Finally, the agent traverses the MST to generate a simple closed path P_1^T, which serves as the coverage trajectory (d).

In the multi-agent learning setting, each sub-region is associated with an evaluation matrix E_i, representing the accessibility of region points for agent i. Based on these matrices, the allocation matrix A is computed as

$$A_{x,y} = \underset{i \in \{1,...,N\}}{\arg\min}\, E_{i|x,y}, \quad \forall (x, y) \in \mathscr{L}, \tag{3.12}$$

where each point (x, y) is assigned to the agent with the lowest evaluation value. Accordingly, the sub-region L_i for agent i is defined as $L_i = \{(x, y) \in \mathscr{L} : A(x, y) = i\}$.

The initial value of E_i is based on spatial distance, formulated as $E_{i|x,y} = d(Pos_i, (x, y)^\tau)$, which favors regions closer to the current position Pos_i of agent i. To adjust the assignment size, a correction factor m_i is introduced and updated via gradient descent:

$$m_i = m_i - \eta(k_i - f)\frac{\partial k_i}{\partial m_i}, \tag{3.13}$$

where k_i is the number of cells currently assigned to agent i, $f = l/N$ is the ideal number of cells per agent, and η is the learning rate.

Finally, a connectivity matrix C_i is applied to emphasize continuous areas near the agent and suppress fragmented regions. The evaluation matrix is refined as $E_i =$

3.5 CPPU Algorithm

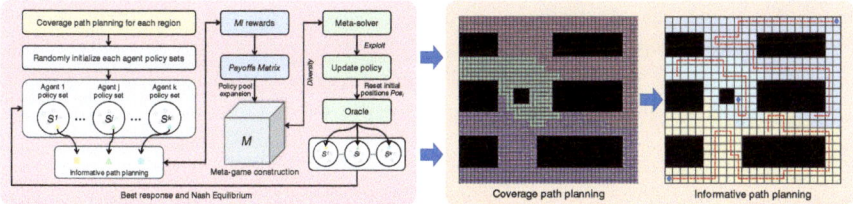

Fig. 3.13 Informative path planning is driven by diverse PSRO

$C_i \odot m_i E_i$. Through this iterative optimization, the algorithm achieves coherent and efficient task allocation for multi-agent coverage.

3.5.5 Optimal CSI Acquisition Path

After generating coverage paths for each region, our goal is to further shorten the path length and determine the optimal path that maximizes effective information. To achieve it, we propose combining the diverse PSRO strategy with the CSI signals, as shown in Fig. 3.13.

Based on our previous research (Zhu et al. 2023), CSI data can be modeled using a MGR model as $N(m(x_i, y_i), \sum_{x_i, y_i})$, where $m(\cdot)$ is the mean function, and $\sum$ is the covariance matrix. The covariance matrix $\sum_{x_i, y_i}$ is computed using the Radial Basis Function (RBF). Let Y_{x_i, y_i} denote the corresponding CSI measurements, whose differential entropy is

$$H\left(Y_{x_i, y_i}\right) = \frac{1}{2} \ln \left|\Sigma_{x_i, y_i}\right| + \frac{n}{2}(1 + \ln(2\pi)). \tag{3.14}$$

The reward based on Mutual Information (MI) is calculated as $MI(Y_{x_i, y_i}; Y_S) = H(Y_{x_i, y_i}) - H(Y_{x_i, y_i} | Y_S)$. Since the differential entropy depends only on the covariance matrix, the reward can be computed analytically without traversing actual paths or performing physical measurements. To estimate the hyperparameters of the RBF and interact dynamically with the environment, a small amount of pilot CSI data (denoted as Y_P) is collected, and the reward is updated by $MI(Y_{x_i, y_i}; Y_S \cup Y_P) = H(Y_{x_i, y_i}) - H(Y_{x_i, y_i} | Y_S \cup Y_P)$.

Next, we use the diverse PSRO algorithm combined with MI-based rewards to explore the informative paths. The algorithm iteratively updates the NE strategies of multiple agents in the overall game and defines the starting positions. For each region, the adaptability of the algorithm is enhanced by processing real-time CSI data. The best response is calculated as

$$BR^i_\varepsilon\left(\pi_\theta^{-i}\right) = \arg\max_{\pi_\theta \in \Delta_{\mathbb{S}^i}} \left[G^i\left(\pi_\theta, \pi_\theta^{-i}\right) + \tau \cdot \text{Diversity}\left(\mathbb{S}^i \cup \{\pi_\theta\}\right)\right],$$

where $\mathbb{S}^i$ is the set of strategies already processed by the i-th agent, and τ is an adjustable constant. The "Diversity" function ensures the diversity of the strategy pool using the Determinantal Point Process (DPP) as $\text{Diversity}(\mathbb{S}) = \mathbb{E}_{\mathbf{Y} \sim \mathcal{P}_\mathbb{S}}[|\mathbf{Y}|] = \text{Tr}\left(\mathbf{I} - (\mathcal{L}_\mathbb{S} + \mathbf{I})^{-1}\right)$. Additionally, exploitability is used to measure the joint strategy as

$$\text{Exploit}(\pi_\theta) = \sum_{i \in N} \left[G^i \left(BR^i \left(\pi_\theta^{-i} \right), \pi_\theta^{-i} \right) - G^i (\pi_\theta) \right]. \quad (3.15)$$

When the value reaches zero, all agents achieve their best responses, constituting a NE.

Finally, an Oracle function is utilized to update the new strategy S_θ, aiming to maximize the utility of each agent while ensuring the diversity of strategies in $\mathbb{S}^i$:

$$\mathcal{O}^1\left(\pi_\theta^2\right) = \arg\max_{\theta \in \mathbb{R}^d} \sum_{S^2 \in \mathbb{S}^2} \pi_\theta^2\left(S^2\right) \cdot \phi\left(S_\theta, S^2\right) + \tau \cdot \text{Diversity}\left(\mathbb{S}^1 \cup \{S_\theta\}\right),$$

where π_θ^2 represents the meta-solver-based strategy of agent 2, and τ is an adjustable constant. Then, we can obtain the efficient information acquisition path planning, and ensure the optimality and diversity of paths.

3.5.6 Updating CSI Data By the GAN

After determining the optimal path $P_{optimal}$, we collect new CSI samples along this path and combine them with pilot CSI data Y_P to predict the CSI distribution at remaining points using a GAN. This approach extends the initial fingerprint database without incurring additional collection costs. GAN is a deep generative model consisting of a generator G and a discriminator D. Through adversarial training, the generator learns to produce realistic data samples p_{data}, while the discriminator attempts to distinguish between real and generated data. The generator G takes a noise vector sampled from a prior distribution $z \in p_z$ as input and generates synthetic samples, while the discriminator D evaluates the likelihood that the input samples come from the real distribution p_{data}. The objective function of GAN is defined as

$$\min_G \max_D V(D, G) = \mathbb{E}_{x \sim p_{data}(x)}[\log D(x)] + \mathbb{E}_{z \sim p_z(z)}[\log(1 - D(G(z)))].$$

The generator G gradually produces samples that closely resemble real data through optimization. We use binary cross-entropy as the loss function by

$$Loss = -y \log \hat{y} - (1 - y) \log(1 - \hat{y}), \quad (3.16)$$

3.5 CPPU Algorithm

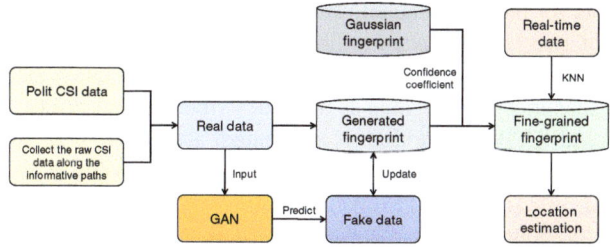

Fig. 3.14 Generating CSI database by GAN and online location estimation

where $y = 1$ and $y = 0$ represent the labels for real and generated data, respectively, and $\hat{y}$ is the output of the discriminator D.

Once trained, the GAN can predict the CSI distribution at remaining sampling points (as shown in Fig. 3.14). We use pilot CSI data Y_P and new CSI data Y_I collected along the optimal path $P_{optimal}$ as real data to train the GAN model for generating virtual data. Through iterative updates, a database CSI_{GAN} is gradually constructed. By combining with the coarse-grained fingerprint database $CSI_{Gaussian}$ built using MGR, the final fine-grained fingerprint database is constructed as

$$CSI_{final} = \varepsilon \times CSI_{GAN} + (1 - \varepsilon) \times CSI_{Gaussian}, \quad (3.17)$$

where ε is a confidence coefficient dependent on the training accuracy of the GAN.

In the online localization phase, the KNN algorithm is employed to estimate the user's position based on real-time CSI measurements. Specifically, the Euclidean distance is computed between the current measurement and each entry in the final fingerprint database CSI_{final}. The K closest fingerprints are selected as neighbors, and a weighted averaging strategy is applied to infer the user's location. The weights are inversely proportional to the distances, ensuring that closer neighbors contribute more significantly to the final estimate. And we use the enhanced fingerprint database, combining the generalization ability of GAN-generated samples and the physical consistency captured by the Gaussian-based coarse database. As a result, the algorithm provides improved localization accuracy while significantly reducing the manual effort required in traditional fingerprint construction. The synergy between offline generative modeling and online matching improves the proposed framework robust and scalable for dynamic indoor environments.

3.5.7 Performance Evaluation

The CPPU algorithm, implemented in Python, extracts information such as graph structure and mutual information rewards, and stores the information-rich paths of agents as inputs for the GAN model. The discriminator and generator of the GAN both consist of two hidden layers (with 64 and 128 neurons, respectively), with a learning rate of 0.0002, trained for 1000 epochs, and a batch size of 32. The experimental scenarios are the same as those described in Sect. 3.4.6.

The proposed CPPU algorithm is evaluated against several state-of-the-art Intelligent Path Planning (IPP) algorithms, including ILUC (Zhu et al. 2024), A3C-IPP (Zhu et al. 2023), DQN-IPP (Wei and Zheng 2022), and GA-IPP (Wei and Zheng 2020). As shown in Table 3.2, the results demonstrate that CPPU consistently outperforms existing methods in both test scenarios. It achieves lower mean errors and standard deviations while significantly reducing the number of sampling points. For example, in Area one, CPPU achieves a mean error of 2.98 m with a standard deviation of 1.55 m, using only 117 sampling points. In Area two, it achieves a mean error of 2.15 m with a standard deviation of 1.12 m, using just 47 sampling points. These results correspond to sampling rate reductions of 63.09 and 73.30%, respectively.

The number of agents directly influences the search path, CSI update accuracy, and the cumulative rewards. As shown in Fig. 3.15, where mutual information and differential entropy are used as reward metrics, increasing the number of agents leads to more efficient exploration and higher total rewards. In Area one, the improvement is evident: with two agents, the reward reaches 151.1 after 10 iterations, while five agents achieve 426.1 after 50 iterations, showing clear benefits in path diversity and sensing quality. In Area two, although the overall reward also increases with more agents, the gap between configurations is smaller. The three-agent setup performs consistently well, showing effective exploration while avoiding excessive coordination overhead. These results indicate that while more agents can enhance performance, the specific environment and the level of coordination required determine the most effective configuration.

To evaluate the robustness and generalizability of the proposed CPPU algorithm, we conducted experiments comparing it with random sampling strategies at rates of 10, 30, and 50%, as shown in Fig. 3.16. The results demonstrate that the CPPU algorithm achieves higher localization accuracy in both test regions, attributed to its adaptive capability to dynamically adjust predictions based on environmental conditions. In contrast, random sampling methods struggle to capture fine spatial variations even at higher sampling rates. Although increasing the sampling rate slightly improves localization accuracy, it significantly raises computational and data collection costs. The CPPU algorithm effectively reduces data acquisition requirements while maintaining high localization performance.

To summarize, the proposed CPPU algorithm demonstrates significant advantages in 6G ISAC by effectively integrating exploration and exploitation within a reinforcement learning framework and enhancing CSI data refinement through GAN-based updates. Experimental results show that CPPU optimizes search trajectories and improves localization accuracy across diverse indoor environments. By employing mutual information and differential entropy as reward metrics, the algorithm adapts to varying scenarios, with multi-agent configurations yielding notably higher cumulative rewards. A deployment of three agents proves to be

3.5 CPPU Algorithm

Table 3.2 Localization experiments and comparisons with the state-of-the-art researches

Algorithms	Area one				Area two			
	Mean errors (m)	Std (m)	Sampling number	Saving sampling rate	Mean errors (m)	Std (m)	Sampling number	Saving sampling rate
CPPU	**2.98**	**1.55**	**117**	**63.09%**	**2.15**	**1.12**	**47**	**73.30%**
Original database	3.22	1.35	317	0%	2.21	1.15	176	0%
ILUC (Zhu et al. 2024)	3.23	1.59	69	78.23%	2.26	1.02	32	81.82%
A3C-IPP (Zhu et al. 2023)	3.70	2.08	69	78.23%	2.30	1.49	32	81.82%
DQN-IPPWei and Zheng (2022)	3.58	1.99	58	81.70%	2.48	1.03	49	72.16%
GA-IPP (Wei and Zheng 2020)	3.78	1.96	72	77.29%	2.58	1.34	57	67.61%

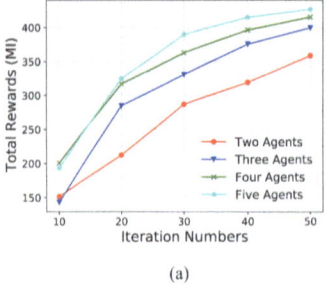 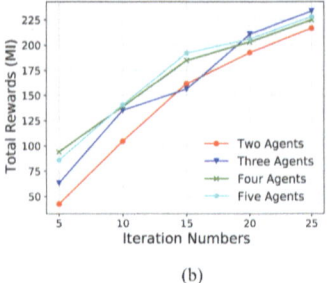

Fig. 3.15 Total reward comparison with different number of agents. (**a**) Area one. (**b**) Area two

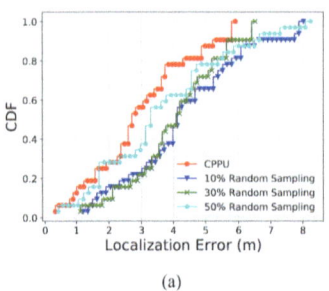 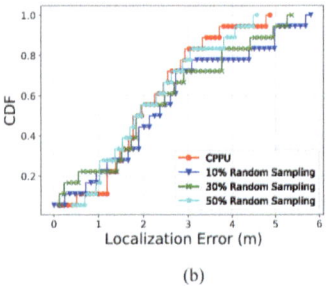

Fig. 3.16 Comparison of localization error with different sampling strategies. (**a**) Area one. (**b**) Area two

the most effective in achieving both efficient coverage and cooperative behavior. Moreover, CPPU exhibits linear scalability with respect to the number of agents and training iterations, substantially reducing the required sampling points and localization errors through the incorporation of generative modeling. Future work will explore advanced techniques for achieving cm-level precision in fingerprint database construction.

References

Faust A, Oslund K, Ramirez O, Francis A, Tapia L, Fiser M, Davidson J (2018) PRM-RL: long-range robotic navigation tasks by combining reinforcement learning and sampling-based planning. In: 2018 IEEE international conference on robotics and automation, pp 5113–5120

Guo J, Wen CK, Jin S, Li GY (2022) Overview of deep learning - based CSI feedback in massive MIMO systems. IEEE Trans Commun 70(12):8017–8045

Halperin D, Greenstein B, Sheth A, Wetherall D (2010) Demystifying 802.11 n power consumption. In: Proceedings of the 2010 international conference on power aware computing and systems, pp 1–5

Liu X, Wu R, Zhang H, Chen Z, Liu Y, Qiu T (2025) Graph temporal convolutional network-based WiFi indoor localization using fine-grained CSI fingerprint. IEEE Sens J 25(5):9019–9033

References

Low ES, Ong P, Cheah KC (2019) Solving the optimal path planning of a mobile robot using improved Q-learning. Robot Auton Syst 115:143–161

Macario Barros A, Michel M, Moline Y, Corre G, Carrel F (2022) A comprehensive survey of visual slam algorithms. Robotics 11(1):24

Wei Y, Zheng R (2020) Informative path planning for mobile sensing with reinforcement learning. In: IEEE conference on computer communications, pp 864–873

Wei Y, Zheng R (2022) A reinforcement learning framework for efficient informative sensing. IEEE Trans Mobile Comput 21(7):2306–2317

Wei Y, Frincu C, Zheng R (2019) Informative path planning for location fingerprint collection. IEEE Trans Netw Sci Eng 7(3):1633–1644

Zheng C, Xu W, Zou Z, Hua T, Yuan C, He D, Zhou B, Liu Z, Lin J, Zhu F, Ren Y, Wang R, Meng F, Zhang F (2025) FAST-LIVO2: fast, direct LiDAR–inertial–visual odometry. IEEE Trans Robot 41:326–346

Zhou P, Wang H, Gravina R, Sun F (2024) WIO-EKF: extended Kalman filtering-based Wi-Fi and inertial odometry fusion method for indoor localization. IEEE Internet Things J 11(13):23592–23603

Zhu X, Qu W, Qiu T, Zhao L, Atiquzzaman M, Wu DO (2020) Indoor intelligent fingerprint-based localization: principles, approaches and challenges. IEEE Commun Surv Tuts 22(4):2634–2657

Zhu X, Qiu T, Qu W, Zhou X, Wang Y, Wu O (2023) Path planning for adaptive CSI map construction with A3C in dynamic environments. IEEE Trans Mobile Comput 22(5):2925–2937

Zhu X, Qiu T, Qu W, Zhou X, Shi T, Xu T (2024) Dynamic radio map construction with minimal manual intervention: a state space model-based approach with imitation learning. IEEE Trans Big Data 11:1799–1812

Open Access This chapter is licensed under the terms of the Creative Commons Attribution 4.0 International License (http://creativecommons.org/licenses/by/4.0/), which permits use, sharing, adaptation, distribution and reproduction in any medium or format, as long as you give appropriate credit to the original author(s) and the source, provide a link to the Creative Commons license and indicate if changes were made.

The images or other third party material in this chapter are included in the chapter's Creative Commons license, unless indicated otherwise in a credit line to the material. If material is not included in the chapter's Creative Commons license and your intended use is not permitted by statutory regulation or exceeds the permitted use, you will need to obtain permission directly from the copyright holder.

Chapter 4
Intelligent Offline Data Updating

Abstract This chapter explores intelligent offline data updating techniques designed to improve the accuracy and robustness of CSI-based localization systems. It begins with an overview of adaptive data sampling strategies, highlighting the differences between traditional and intelligent updating approaches and identifying the key challenges in maintaining real-time fingerprint reliability. Next, we introduce CSI prediction models, including machine learning-based approaches, hybrid methods that combine real and predicted data, as well as techniques based on crowdsourcing and multivariate Gaussian regression. To address missing CSI data, we investigate various generative strategies, including the application of large-scale models for data generation. We also evaluate robustness enhancements achieved through data augmentation and discuss the limitations with synthetic data.

In addition to prediction and augmentation strategies, we further investigate methodologies for constructing and refining CSI fingerprint databases. This includes building initial radio maps, analyzing CSI error bounds, and proposing behavior cloning techniques based on imitation learning for fine-grained radio map generation. The chapter concludes by introducing the Deep-Broad Learning (DBLG) algorithm, outlining its motivation, system architecture, the integration of a GAN model with confidence-based weighting, and experimental validation of its performance.

Keywords Fingerprint update · Generative models · Radio map

4.1 Overview of Offline Data Updating Techniques

4.1.1 Adaptive Data Sampling Techniques

In CSI-based indoor localization systems, offline data collection serves as the foundation for constructing fingerprint databases. Static sampling strategies often result in inefficiencies, as they fail to account for spatial variability and environmental complexity. Adaptive data sampling techniques address these limitations by dynamically adjusting sampling density and position based on signal characteristics (Zheng et al. 2019). Compared to uniform or manual sampling, adaptive methods improve coverage efficiency while minimizing redundant measurements. They are particularly well-suited for offline stages, where data acquisition is time-consuming and resource-intensive, and where representative sampling can significantly affect the accuracy of subsequent localization models.

Deep reinforcement learning offers an effective approach for optimizing data sampling strategies in offline acquisition. By treating the sampling process as a decision-making problem, the model can learn to adjust sampling behavior based on observed environmental feedback. Reward signals such as localization accuracy or coverage completeness guide the agent to focus on high-variability regions and reduce sampling in stable areas. Through iterative training, the model develops policies that support more efficient data collection across large and complex indoor spaces. These learned policies are capable of adapting to diverse environmental configurations without the need for manual intervention. Moreover, deep reinforcement learning models can incorporate long-term rewards, allowing the agent to anticipate the overall impact of sampling decisions rather than relying solely on local feedback. This further improves sampling effectiveness and reduces unnecessary effort during the offline stage.

Transfer learning also enhances offline sampling efficiency by leveraging knowledge from previously explored environments. Sampling policies or signal feature extractors trained in one building or room can be adapted to similar locations with minimal fine-tuning. This reduces the need for exhaustive data collection in every new setting and accelerates the deployment of localization systems across different sites. For example, a model that has learned signal behavior patterns in one office layout can guide sampling decisions in another area with comparable structural characteristics. In practice, pre-trained models can retain generalizable features in early network layers while adapting to domain-specific variations through updates to deeper layers. This selective reuse of learned representations enables robust performance even when environmental changes are present. The ability to reuse knowledge across domains not only enhances scalability but also significantly reduces computational and labor costs during the fingerprint database construction phase.

4.1.2 Traditional Methods vs. Intelligent Updating Techniques

Traditional offline data updating methods primarily rely on periodic data re-collection and manual annotation. While these approaches can maintain model usability over time, they incur significant time and labor costs and struggle to adapt to real-time environmental changes. As indoor environments evolve, traditional methods often fail to reflect these changes promptly, leading to reduced system performance. In contrast, intelligent updating techniques leverage machine learning

4.1 Overview of Offline Data Updating Techniques

and optimization algorithms, offering a more efficient and adaptive solution for handling dynamic and complex indoor environments (Li et al. 2022).

Traditional approaches exhibit several key limitations that hinder their effectiveness in rapidly changing environments. Periodic data recollection, often performed at fixed intervals, fails to promptly respond to sudden changes in the indoor environment, such as alterations in furniture layout or unexpected signal interference. This delay can result in data latency, which degrades the accuracy of the system. Moreover, the manual process of annotating and correcting data is not only labor-intensive and time-consuming but also unsustainable for large-scale deployments. It also increases operational costs and can introduce subjective errors, further compromising the system's reliability. Additionally, traditional models, trained using static data, lack self-adjustment capabilities. When environmental changes occur, these models necessitate a complete recollection of data and a full retraining process, which is both resource-intensive and time-consuming, making it difficult to meet real-time system requirements. Traditional methods also assume gradual environmental changes, rendering them inadequate for rapidly changing conditions. As a result, static update strategies often fail to address scenarios such as sudden furniture rearrangement or increased wireless interference, leading to performance degradation.

Intelligent updating techniques, on the other hand, provide significant improvements by utilizing advanced machine learning algorithms such as reinforcement learning, transfer learning, and self-supervised learning. These techniques enable more efficient and adaptive data updates, overcoming many of the limitations of traditional methods. Reinforcement learning, for instance, allows the system to dynamically adjust its data collection strategies and model parameters based on continuous environmental feedback (Wong et al. 2023). This ensures that updates are triggered only when necessary, thereby reducing resource waste and computational overhead. Transfer learning, by contrast, enables the system to leverage pre-trained models from existing environments and fine-tune them with minimal new data, thereby minimizing the need for exhaustive data recollection and manual annotation (Guo et al. 2023). This significantly cuts down on update costs and the time required for model adaptation. Additionally, these intelligent techniques can predict environmental trends based on historical data, proactively adjusting model parameters to maintain high positioning accuracy even during environmental fluctuations. Unlike traditional methods that often require full-scale updates, intelligent techniques adopt incremental updates and online learning approaches. These strategies modify only the necessary components of the model, which dramatically reduces computational and storage costs while enhancing operational efficiency.

To summarize, intelligent updating techniques address the key shortcomings of traditional methods by incorporating advanced machine learning approaches, as listed in Table 4.1. These techniques offer significant improvements in efficiency, adaptability, and cost-effectiveness, making them an essential component in maintaining the long-term stability and accuracy of indoor localization systems. By minimizing unnecessary data collection and enhancing the system's ability to respond to dynamic changes, intelligent methods support a more scalable and reliable framework for the continuous evolution of localization systems.

Table 4.1 Comparison of traditional methods and intelligent update techniques

Aspect	Traditional methods	Intelligent update techniques
Data collection	Periodic re-collection at fixed intervals, requiring manual data labeling and extensive human effort. Limited by environmental changes and human inconsistency.	Dynamic collection using machine learning algorithms to adjust sampling density based on real-time environmental changes. Adaptive and focused data collection, minimizing redundant efforts.
Adaptability	Limited. Requires full re-collection and model retraining to accommodate major environmental changes, leading to significant downtime.	Highly adaptive. Machine learning models automatically adjust based on real-time feedback, allowing for quick adaptation to dynamic changes in the environment.
Cost	High. Significant labor and time costs associated with manual data collection, annotation, and retraining. Operational costs escalate as the system scales.	Lower. Automation reduces labor costs. Pre-trained models and fine-tuning allow for reduced reliance on large-scale data collection, decreasing overall cost.
Time efficiency	Low. Data re-collection and model retraining are time-consuming, with significant delays in system updates.	High. Incremental updates using pre-trained models and online learning reduce update time, providing faster deployment and real-time adaptability.
Flexibility	Inflexible. Cannot quickly adapt to sudden environmental changes; requires complete retraining for new settings.	Flexible. Self-adjusting algorithms allow for efficient handling of unexpected environmental changes without full retraining.
Scalability	Difficult to scale for large environments. As the system grows, so does the manual labor required, limiting scalability.	Easily scalable. Models can be transferred and adapted to different environments, reducing the need for extensive retraining and enabling deployment in large-scale or multi-site environments.
System robustness	Vulnerable to sudden environmental fluctuations such as structural changes or new interference sources. Leads to performance degradation.	Robust. Proactive machine learning algorithms adjust in response to sudden environmental changes, maintaining system stability and high performance.
Model retraining	Full retraining is necessary when environmental conditions change significantly, which can be resource-intensive and time-consuming.	Incremental updates. Models can adapt to new data with minimal retraining, significantly reducing computational and time costs.

4.1.3 Challenges in Real-Time Fingerprint Maintenance

In real-world deployments, maintaining a fingerprint database in real time presents several critical challenges. These include rapid environmental changes, limited computational resources, and difficulties in ensuring data quality. Such challenges affect not only the timeliness and accuracy of indoor positioning but also impose higher requirements on computational efficiency and system scalability (Merenda et al. 2022). This chapter analyzes these challenges in detail from three perspectives: **environmental dynamics**, **system resource constraints**, and **data consistency**.

Environmental dynamics introduce considerable instability to the fingerprint database. Indoor wireless signal propagation is highly sensitive to environmental changes, and even minor variations may lead to outdated fingerprints and degraded localization performance. For instance, fluctuations in human crowd density cause noticeable shifts in signal patterns between peak and off-peak periods. Modifications to the physical layout, such as furniture rearrangement or partition installation, can significantly alter signal paths and distributions, requiring the system to adapt quickly to updated signal characteristics. In addition, wireless interference from devices like Bluetooth equipment and additional routers introduces random and unpredictable fluctuations, especially in environments with dense deployments. These dynamic factors make real-time fingerprint maintenance more complex and reduce the reliability of static data representations.

Computational resource limitations further complicate real-time database updates. In large-scale environments, continuous data collection and frequent model updates generate heavy computational and storage demands. Traditional approaches often involve full model retraining, which is time-consuming and resource-intensive. As coverage areas expand and signal conditions become more diverse, the volume of fingerprint data increases sharply, making it more difficult to maintain responsiveness and meet real-time operational requirements. Furthermore, in applications that require long-term storage of historical fingerprints, storage overhead grows rapidly and can become a performance bottleneck during real-time updates.

Data quality management also poses significant challenges. Environmental variability and computational constraints can lead to issues such as inconsistent data, noise interference, and redundant information. Data collected at different times may exhibit substantial variation due to changing surroundings, and if not processed promptly, such inconsistencies can lower the reliability of the fingerprint database. Wireless signals are vulnerable to various types of interference, including device noise and multipath propagation, and unfiltered noisy data can distort training outcomes. In scenarios involving frequent updates, large amounts of redundant data may accumulate, increasing both storage requirements and training complexity without contributing to improved model performance.

4.2 Predictive Models for CSI Fingerprint Updates

4.2.1 Machine Learning Models for Predicting CSI Changes

To maintain positioning accuracy in dynamic indoor environments, it is essential to address the rapid obsolescence of CSI fingerprint data. Environmental changes such as human movement, signal interference, and structural adjustments often lead to significant fluctuations in wireless signal characteristics. Static update strategies are typically unable to respond in a timely manner, which highlights the necessity of predictive models that can anticipate CSI changes and guide intelligent offline updates (Zhang et al. 2021).

ML algorithms offer a powerful toolkit for forecasting CSI variations by learning from historical signal patterns (Zhang et al. 2021). These models help reduce the need for frequent manual data collection and enable systems to remain accurate and responsive in changing conditions. By anticipating environmental trends, the update process becomes more efficient and cost-effective. Moreover, predictive modeling enables selective updates, where only the most affected regions of the fingerprint database are refreshed, thereby avoiding unnecessary computation and reducing storage overhead. This targeted approach improves system responsiveness and is especially advantageous in large-scale deployments where full updates are impractical. Additionally, predictive models can serve as an early warning mechanism, flagging areas where signal drift is likely to occur, which supports proactive resource allocation and system reconfiguration.

Predictive models for CSI changes can be broadly classified into three main types. Time series models such as Autoregressive, Autoregressive Moving Average (ARMA), and LSTM networks are effective at capturing temporal dependencies within signal sequences, making them suitable for modeling both short-term fluctuations and long-term trends. Regression-based methods, including Support Vector Regression, Random Forest Regression, and Broad Learning Systems, are adept at learning high-dimensional feature relationships from past observations and can handle nonlinear dependencies with high accuracy. Generative models like Variational Autoencoders (VAE) and GAN are capable of synthesizing realistic future CSI samples, which can be used to enrich datasets and support updates in data-sparse conditions.

The benefits of these predictive models are manifold. They capture both temporal and spatial signal correlations, reduce the overhead of repeated data acquisition, and improve system robustness. Integrating reinforcement learning techniques further enables dynamic adjustment of model parameters based on environmental feedback, enhancing adaptability. However, these models are not without limitations. Time series models may struggle with highly nonlinear dynamics, while generative models often require significant computational resources for training and inference. Future research may address these issues through model compression, multi-source data fusion, and distributed training frameworks to improve scalability and deployment efficiency in real-world applications (Jianhua et al. 2024).

4.2.2 Hybrid Models Combining Real and Predicted Data

For CSI fingerprint updates, relying on either real data or predicted data presents inherent limitations. Real data, while offering high accuracy, requires extensive collection efforts and cannot always reflect the full range of dynamic changes within indoor environments. However, predicted data can compensate for incomplete measurements and reduce the need for large-scale data collection, but its reliability is constrained by model performance and may be affected by uncertainties in complex or rapidly changing conditions. To address the above challenges, hybrid models that integrate real and predicted data have become a practical and effective solution for maintaining and updating CSI fingerprints.

The core principle of hybrid models is to combine a limited set of real measurements with predicted data to support dynamic and adaptive fingerprint updates, thereby improving both accuracy and spatial coverage. In practical applications, the system initially collects a subset of real CSI data through deployed sensor networks or user devices, which serve as reliable reference samples. Based on these samples, machine learning models or statistical regression methods such as multivariate Gaussian regression or DNN are employed to estimate CSI characteristics in regions that lack direct measurements. These predictions are derived through interpolation or extrapolation techniques, enabling the estimation of signal features in areas where real data is sparse or unavailable. Once both data sources are available, techniques like weighted data fusion, Bayesian inference, or Kalman filtering are applied to integrate the real and predicted values. The process aims to preserve the accuracy of real measurements while benefiting from the extended spatial coverage provided by the predicted data, ultimately resulting in a more complete and consistent fingerprint database (Huan et al. 2022).

Hybrid models offer several practical advantages. First, they significantly reduce the cost and workload associated with data collection by minimizing the amount of real data required, which is especially valuable in large or complex indoor environments. Second, by filling in signal-blind zones or physically inaccessible areas with predicted data, these models enhance the overall coverage and continuity of the fingerprint database. It ensures that the licalization system can function effectively even in previously unmeasurable regions. Third, the combination of real measurements enables the correction of predicted values, thereby improving the overall reliability and robustness of fingerprint updates. The ability to adjust and refine predictions using real data allows the system to maintain high localization accuracy and consistency for the environmental changes. As a result, hybrid models provide an efficient approach to CSI fingerprint maintenance, capable of supporting stable and precise indoor positioning services across a wide range of deployment scenarios.

4.2.3 Crowdsourcing and Multiple Gaussian Regression

Timely and comprehensive CSI fingerprint updates are essential for maintaining positioning accuracy in dynamic indoor environments. Traditional static fingerprint databases often fail to adapt to environmental changes, leading to reduced system performance over time. To address this, two prominent approaches have been widely explored: crowdsourcing-based data collection and MGR-based data prediction. Crowdsourcing utilizes the sensing capabilities of user devices to collect real CSI samples during regular usage, while MGR estimates unobserved CSI values through spatial and temporal modeling (Wei and Zheng 2021; Le et al. 2021). These methods offer distinct advantages and challenges in terms of coverage, data quality, computational complexity, and adaptability.

Crowdsourcing enables large-scale CSI data collection by leveraging user devices such as smartphones, wearables, and IoT nodes. During routine network interactions, these devices capture CSI features, including amplitude, phase, and timestamps. Data is uploaded to a central server where it undergoes filtering and preprocessing to remove noise, device-induced variability, and outliers. Given the diversity and heterogeneity of data sources, techniques such as anomaly detection and time-weighted integration are employed to ensure data quality and prioritize recent measurements. By continuously incorporating fresh samples from multiple users, the fingerprint database can be incrementally updated, improving both spatial coverage and temporal relevance.

In contrast, MGR models aim to predict CSI values in regions where data is sparse or unavailable. The approach assumes that CSI features follow a Gaussian process governed by spatial and temporal correlations. A covariance function, such as the radial basis function or polynomial kernel, is selected to quantify similarity between measurement points. Model training involves estimating hyperparameters using maximum likelihood estimation based on available CSI samples. Once trained, the model can interpolate missing data points or forecast future CSI values, along with associated uncertainty estimates. It enables proactive fingerprint updates in areas where real measurements are difficult to obtain, such as restricted zones or rarely visited regions. And a comparative overview of these two approaches is presented in Table 4.2.

These two approaches are complementary in nature. Crowdsourcing provides scalable and up-to-date data through passive collection, which is particularly effective in high-traffic environments. However, it depends heavily on user participation and suffers from inconsistency across devices. MGR offers reliable estimations in data-sparse regions by leveraging spatial and temporal models, but it lacks real-time responsiveness and demands high computational resources. Combing both methods into a unified fingerprint maintenance framework allows the system to benefit from the strengths of each, and using crowdsourced data for frequent updates and MGR to complete and correct the fingerprint map where real data is lacking. And it enables the construction of a robust, efficient, and adaptable CSI fingerprint database suitable for dynamic indoor environments.

Table 4.2 Comparisons of crowdsourcing and MGR

Aspect	Crowdsourcing	MGR
Core mechanism	Collects real-time CSI data from user devices during daily activities. Relies on large-scale user participation to generate continuous updates.	Uses Gaussian process models to estimate missing or outdated CSI values through spatial-temporal interpolation based on existing data.
Data source	Real CSI measurements from heterogeneous Wi-Fi devices, with variations due to hardware diversity and user behavior.	Historical CSI data from controlled environments, used as training input for predictive modeling.
Spatial coverage	Broad. Capable of covering diverse environments due to user mobility and device diversity. Especially effective in high-traffic zones.	Limited by the scope of training data. Performance may degrade in previously unmeasured or rapidly changing areas.
Update timeliness	Near real-time updates enabled by continuous data upload from active users. Suitable for dynamic environments.	Depends on model retraining frequency. Less responsive to real-time environmental changes.
Data quality	May be affected by noise, device inconsistencies, and user movement. Requires robust preprocessing and outlier detection.	Generates smooth and consistent outputs, but predictions may deviate if underlying assumptions are violated.
Accuracy	High in areas with dense user activity and frequent data uploads. Performance decreases in sparse regions.	Effective for filling missing data and smoothing fingerprint distributions in sparse regions. Accuracy depends on kernel design and training quality.
Computational requirements	Low. Preprocessing and integration can be performed on edge devices or lightweight cloud services.	High. Requires centralized training and inference infrastructure, especially for large-scale datasets.

4.3 Techniques for Simulating Missing CSI Data

4.3.1 The Application of Large-Scale Models in Data Generation

In real-world indoor localization systems, missing CSI is a frequent and critical issue, typically caused by hardware malfunctions, unstable wireless channels, environmental occlusion, or packet transmission errors. These data gaps reduce the completeness and reliability of the CSI fingerprint database, thereby affecting downstream tasks such as location inference and environment-aware decision-making. Traditional imputation methods often struggle to capture the complex spatial and temporal dependencies embedded in CSI data. In response, large-scale generative models have emerged as a promising solution for simulating and

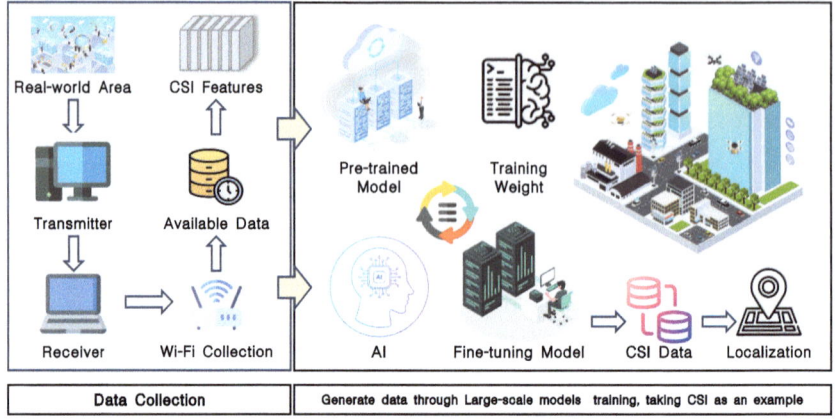

Fig. 4.1 The application of large-scale models in data generation

supplementing missing CSI, offering improved adaptability, generalization, and efficiency across various deployment scenarios (Zhu et al. 2025).

As shown in Fig. 4.1, the overall workflow begins with data acquisition from a physical environment using Wi-Fi transmitters and receivers, followed by CSI feature extraction and dataset construction. The available data is then used to fine-tune powerful pre-trained generative models, which can produce synthetic CSI data that mimics real signal patterns. These models support a variety of advanced functions: predicting missing or corrupted entries, generating realistic new samples to augment training datasets, filtering out environmental noise to refine signal quality, and extrapolating future CSI states based on historical temporal sequences. These capabilities not only improve the robustness of localization systems but also significantly reduce the cost and effort of data collection in large-scale deployments.

The ability of large-scale models to handle CSI data generation is underpinned by major advances in generative deep learning, including ***transformer-based architectures, diffusion models, and hybrid frameworks that combine data-driven and physics-aware learning.*** The continuous growth of CSI training datasets, along with improvements in GPU and cloud computing infrastructure, has made it possible to generate complex and diverse CSI distributions with high precision. More importantly, by embedding physical constraints related to signal propagation into the modeling process, these generative models can better align with real-world wireless behavior. Looking forward, the development of lightweight and energy-efficient architectures will support deployment on edge devices, while multimodal fusion involving vision, LiDAR, or radar data is expected to further enrich CSI signal modeling.

4.3.2 Data Augmentation for Enhanced Model Robustness

The quality and completeness of CSI data are critical for the performance of indoor localization systems. However, practical deployments often encounter challenges such as hardware-induced noise, dynamic environmental changes, and instability during data acquisition, all of which may result in missing values or outliers. To mitigate these issues without incurring additional data collection costs, data augmentation offers an effective strategy for improving model robustness and generalization. For CSI data, augmentation techniques must be carefully designed to maintain the underlying physical properties of wireless signals.

To tackle challenges such as incomplete measurements and signal distortions, a variety of data augmentation techniques have been developed specifically for CSI-based indoor localization. These techniques aim to simulate realistic channel variations while preserving the structural and statistical characteristics of wireless propagation. Representative strategies include:

- **Noise injection**: Introducing Gaussian or Poisson noise into raw CSI data to emulate fluctuations caused by diverse environmental interference.
- **Time-series smoothing**: Applying moving average or exponential smoothing techniques to reduce burst noise and enhance temporal stability in CSI sequences.
- **Signal interpolation**: Using linear or spline interpolation to recover missing CSI values, thereby restoring the continuity of spatial fingerprints.
- **Random channel masking**: Randomly obscuring certain subcarriers or antennas during training to improve model robustness against partial channel data loss.
- **Data mixing**: Generating new samples by combining multiple CSI inputs through weighted averaging or nonlinear transformations, increasing diversity in the training set.

These techniques have proven effective across a wide range of CSI-based tasks. In deep learning-driven localization, augmented CSI data enable models to generalize more effectively across different spatial environments. In semi-supervised and self-supervised learning settings, augmentation facilitates the exploitation of unlabeled data, enabling efficient training even with limited annotations. And we believe that combining data augmentation with adversarial learning and self-supervised frameworks is expected to further boost model adaptability in complex, dynamic environments. Additionally, automated augmentation strategies such as AutoAugment and RandAugment hold promise for discovering optimal augmentation policies tailored to CSI-specific characteristics.

4.3.3 Synthetic Data Generation and Its Limitations

As machine learning becomes increasingly necessary to CSI-based indoor localization systems, synthetic data generation has become a valuable means of addressing challenges such as incomplete measurements, limited labeled datasets, and variations across deployment scenarios. By supplementing real-world data with synthetic samples, models can achieve improved generalization and robustness while reducing the costs and constraints of large-scale data acquisition. When properly designed, synthetic CSI data can support more flexible system development and faster deployment across heterogeneous environments (Suroso et al. 2022).

Multiple techniques are currently employed to generate synthetic CSI data. Statistical model-based methods produce samples by modeling the underlying CSI distribution using multivariate Gaussian or Markov processes, aiming to replicate realistic variations in signal behavior. GAN learns to generate high-fidelity CSI samples through adversarial optimization, capturing complex data distributions from real-world environments. VAE encodes CSI into a latent space and generates new samples through learned decoding functions, allowing for smooth variation and interpolation between signal states. Physical simulation methods rely on established channel models, such as Rayleigh or Rician fading, to produce CSI samples that adhere to known wireless propagation characteristics.

While synthetic data generation offers advantages such as controllable parameter tuning and reduced dependency on hardware, it also presents several limitations in practice. Many statistical or learning-based methods make assumptions that oversimplify real-world signal behavior, resulting in a mismatch between synthetic and actual data distributions. Artifacts or inconsistencies may be introduced during generation, particularly in GAN or VAE outputs, which could impair model learning if not properly filtered. Furthermore, physical simulation often fails to fully capture the dynamic and cluttered nature of indoor environments, especially in scenarios with multipath propagation and human interference. These issues can lead to decreased accuracy and model overfitting when synthetic data are used without careful validation or adaptation.

Despite these challenges, synthetic data has been effectively applied in CSI-based localization tasks, particularly for pre-training deep models, augmenting limited datasets, and simulating domain-specific signal variations. When used in combination with real data, synthetic samples can enhance model resilience to environmental changes and reduce reliance on manual data collection, thereby supporting more scalable and efficient system development.

4.4 Generate CSI Fingerprint Algorithm

4.4.1 Preliminaries

Utilizing crowdsourcing techniques and GPR modeling offers an effective solution to reduce the manual workload during the offline phase. Crowdsourcing enables the autonomous construction of fingerprint databases by passively collecting data from users' daily activities without explicit effort, thereby significantly saving labor, material, and time costs. Furthermore, integrating Bayesian inference methods allows for dynamic fingerprint data perception, facilitating both the initial construction and continuous updates of fingerprint databases (Wang et al. 2011).

However, sensed fingerprint data inevitably contain noise, which undermines the construction of high-precision fingerprint databases. While robotic platforms can optimize data collection paths (Wei et al. 2019), this approach faces several limitations. First, due to complex indoor environmental layouts, robotic agents often fail to achieve full coverage during path exploration. Second, methods like multivariate Gaussian regression and greedy algorithms demand substantial computational resources and are prone to local optima. To advance the applicability of fingerprint-based localization and deliver high-quality location services, particularly for fine-grained fingerprint database construction, we propose a CSI fingerprint update algorithm based on imitation learning. Extensive real-world experiments validate the effectiveness of the proposed method, demonstrating an 18.2% improvement in localization accuracy.

4.4.2 The Overview

The proposed algorithm integrates three core modules: multivariate Gaussian regression, imitation learning, and path planning, as shown in Fig. 4.2. Through three rounds of iterative refinement, it autonomously constructs a fine-grained fingerprint database. Unlike conventional approaches requiring exhaustive offline sampling, our method significantly reduces manual labor while maintaining high fingerprinting accuracy. For precise localization, we employ a KNN-enhanced CSI positioning mechanism that dynamically adapts to environmental variations.

In the offline phase, we first randomly select sampling points (following normal distribution) to collect raw CSI fingerprint data and construct a coarse-grained initial fingerprint database using multivariate Gaussian regression. Concurrently, the A3C-IPP algorithm is employed to grid the target area into a vertex-edge graph for optimal path planning that maximizes information gain. New CSI data collected along this path are used to compute error bounds through state-space modeling, which then calibrates the initial fingerprint database. Subsequently, imitation learning extracts CSI features to train prediction models for estimating spatial distributions at remaining points. Finally, confidence coefficients derived

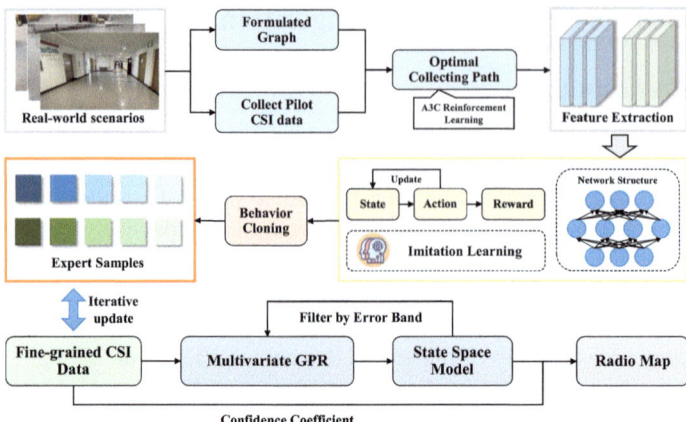

Fig. 4.2 The overall of the proposed algorithm

from imitation learning balance predictions with raw measurements, yielding a refined fine-grained fingerprint database.

In the online phase, the system employs a KNN-based fingerprint-matching algorithm for user location estimation. Unlike conventional approaches that rely on continuous CSI data packets, our proposed solution constructs a lightweight positioning framework by extracting the core features of CSI signals. This design not only eliminates the dependency on high-density sampling but also simplifies complex CSI data processing into an RSS-like fingerprint recognition paradigm. While maintaining positioning accuracy, the method significantly reduces computational complexity.

4.4.3 Constructing the Initial Radio Map

This chapter describes the methodology for fitting a coarse-grained initial fingerprint database using MGR and constructing an optimal path for fingerprint collection. Following conventional approaches, the target localization area is first partitioned into grids, where adjacent sampling points are interconnected to form a fingerprint map. Using the data acquisition equipment mentioned previously, approximately 10% of the total sampling points are randomly selected based on a normal distribution for the first round of CSI data collection.

Under ideal conditions, data collected by the same device should exhibit a Gaussian distribution in spatial domain. However, due to complex indoor environmental factors and noise interference, the relationship between physical coordinates **x** and

4.4 Generate CSI Fingerprint Algorithm

raw CSI measurements **y** follows a multivariate Gaussian regression distribution, expressed as

$$y \mid x \sim N\left(m(x), \sigma_f^2\right), \tag{4.1}$$

where $m(\mathbf{x})$ denotes the mean function and σ_f^2 represents the covariance function. A RBF is employed as the kernel to compute σ_f^2 as

$$\sigma_f^2 = \exp\left(-\frac{1}{2\tau^2}\|x - x'\|^2\right), \tag{4.2}$$

where τ denotes the scaling coefficient. Subsequently, after fitting the model with the initial dataset, we can predict the CSI distribution at remaining sampling points. The fingerprint data is then updated using a Kalman filter to generate the initial fingerprint map.

The initial 10% sampled data proves insufficient for constructing high-precision fingerprint models. To enhance database performance, additional CSI data collection becomes necessary. We employ the A3C reinforcement learning algorithm (detailed in previous Sect. 3.4.4) to identify optimal sampling paths. The proposed algorithm, built upon an Actor-Critic framework with multithreading technology, demonstrates superior performance compared to conventional Q-learning approaches. The agent generates new sampling trajectories based on the initial fingerprint map, significantly reducing manual collection costs. The overall training objective is

$$\theta = \theta + \alpha \nabla_\theta \log \pi_\theta(s_t, a_t) A + \beta \nabla_\theta H(\pi(s_t, \theta)), \tag{4.3}$$

where θ is the parameters of π_θ, s_t denotes the agent's positional state with an action space containing {up, down, left, right}, and α, β are scaling parameters. The exploration path is set between the diagonal start and end points, allowing the agent to search for one or multiple paths within limited steps while satisfying

$$Path_{optimal} = \arg\max_{Path \in \Psi} R(Path), \tag{4.4}$$

where $Path$ represents a potentially valid path, $R(\cdot)$ denotes the reward function, and Ψ encompasses the complete set of paths. We select the path with the maximum total reward value from this collection and designate it as the final optimal path to guide subsequent data acquisition strategies.

4.4.4 The Error Band of CSI

The coarse-grained initial radio map constructed from limited raw CSI data fails to meet high-precision requirements, prompting further optimization using state-space models. The CSI data collection process can be analogized to a Markov model. State-space models establish connections between observable variables and internal system states. When treating the CSI sequence as a first-order integrated series with stationary differences, it can be fitted using a model containing autoregressive lag terms and other parameters. The specific model is as follows

$$\Delta y_t = c + \phi_1 \Delta y_{t-1} + \theta_1 \epsilon_{t-1} + \epsilon_t, \tag{4.5}$$

where c represents the intercept of the ARMA model, Δ denotes the first-order difference operator, and ε_t follows a normal distribution with mean 0 and variance σ^2. We employ the statsmodels toolkit to fit the pilot data and subsequently predict the error range of raw CSI samples.

Subsequently, the Savitzky-Golay algorithm is applied for data smoothing. This algorithm enhances data precision without altering signal characteristics through a set of "convolution coefficients" implemented via linear least squares fitting

$$x_{k,smooth} = \bar{x}_k = \frac{1}{H} \sum_{i=-w}^{+w} x_{k+i} h_i, \tag{4.6}$$

where h_i represents smoothing coefficients, here set to the error range. This completes the second-stage filtering of the initial radio map $Map_{initial}$, yielding the refined result Map_{second}.

4.4.5 Behavior Cloning of CSI Based on Imitation Learning

This chapter describes the use of imitation learning for behavior cloning on raw CSI fingerprint data. Imitation learning demonstrates strong performance in sequential decision-making problems by learning from expert demonstrations, enabling the model to generate state-action trajectory distributions that match the distribution of the input trajectories. The supervised learning process, referred to as behavior cloning, seeks to find an optimal policy π_θ with parameters θ that solves the following optimization problem:

$$\theta^* = \arg\max_{\theta} \mathbb{E}_{\rho \in D} [p_\theta (a_{1:T} \mid s_{1:T})], \tag{4.7}$$

where $\pi_0(a|s)$ is the probability of executing action a under state s according to the optimal policy. Behavior cloning satisfies $p_0(a_{1:t}|s_{1:t}) = \prod_{t=1}^{T} \pi_0(a_t|s_t)$.

4.4 Generate CSI Fingerprint Algorithm

Imitation learning achieves data processing through behavior cloning from expert demonstrations. However, obtaining precise samples is challenging, and trajectory errors accumulate over time,

$$E[errors] \leq \varepsilon(T + (T-1) + (T-2) + \ldots + 1) \propto \varepsilon T^2, \tag{4.8}$$

where ϵ denotes the error probability at time step t. The proposed algorithm employs imitation learning to accurately predict and update the CSI data distribution, decomposing the sample CSI sequence segments and mapping them to high-level codes for input sequence reconstruction.

The training trajectory p is partitioned into N segments $[\text{seg}_1, \text{seg}_2, \ldots, \text{seg}_N]$, where

$$\text{seg}_i = \left[(s_{b_{i'}}, a_{b_{i'}}), (s_{b_{i'}+1}, a_{b_{i'}+1}), \ldots, (s_{b_i-1}, a_{b_i-1})\right], \tag{4.9}$$

where $i' = i - 1$. The sample behavior cloning is transformed into learning sub-tasks to generate a policy $\pi_0(a|s, l)$ (l being the latent variable). Using autoencoders, imitation learning is performed for each sub-task to obtain specific policies (Kipf et al. 2019). Disjoint segments are recombined to form new samples similar to data augmentation, where all sub-policies can be mapped to the training trajectory.

The segment l can be encoded as

$$p_\theta(a_{1:T} \mid s_{1:T}) = \sum_{b_{1:N}} \sum_{l_{1:N}} p_\theta(a_{1:T} \mid s_{1:T}, b_{1:N}, l_{1:N}) \, p(b_{1:N}, l_{1:N}), \tag{4.10}$$

It can be decomposed across time steps, where single-segment decomposition prevents over-writing of training trajectories. Then, we construct the recognition model as

$$q_\phi(b_{1:N}, l_{1:N} \mid x_{1:T}) = \prod_{i=1:N} q_{\phi_l}\left(l_i \mid x_{b_i'} : b_i - 1\right) q_{\phi_b}\left(b_i \mid x_{b_{i'}:T}\right). \tag{4.11}$$

Parameters are shared across segments, with networks $q_{\phi_l}(l|x)$ and $q_{\phi_\beta}(b|x)$ as the core components. The model employs LSTM and MLP to predict the logical relationships at the final step of sub-task seg_i, producing an n-dimensional vector containing class information, which serves as the encoder. This encoder only predicts states, and the lower bound of sub-segments may reduce precision. As shown in Fig. 4.3, the original CSI data is used as training trajectories. The encoder computes $q(b_i|x)$, obtains sub-segment seg_i from $q(l_i|x)$, stores the reconstruction sample loss in a buffer, and finally derives the policy $\pi_0(a_t|s_t, l_i)$ through imitation learning via the decoder.

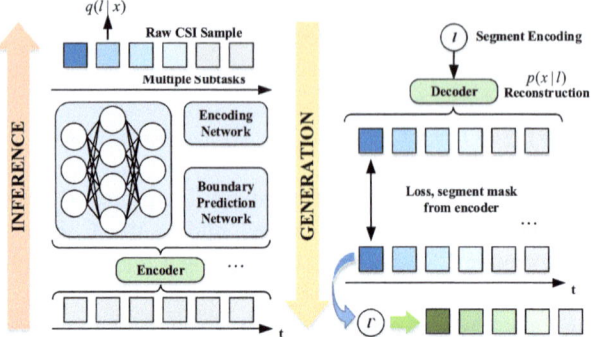

Fig. 4.3 Subtasks for a training trajectory

4.4.6 The Fine-Grained Radio Map

CSI data are complex-valued and stored as *Tensor* objects. While imitation learning can update the real components of samples, it fails to update the imaginary components. Moreover, models trained with limited real-world data can only approximate real-valued quantities. Therefore, it is necessary to balance the relationship between updated data and ground truth values, and develop methods for accurate inference of imaginary components (Huai et al. 2023).

In the above chapter, the *batch_acc* computed through imitation learning serves as an effective evaluation metric, denoted as the confidence coefficient ϵ. It represents the cosine similarity between updated data and ground truth values, i.e., the regression accuracy. The algorithm combines a MGR model with imitation learning for third-round iterative correction, as expressed in the following

$$Map_{final} = \varepsilon \times Map_{second} + (1-\varepsilon) \times Map'_{second}. \tag{4.12}$$

From an information validity perspective, the more raw CSI samples collected, the more precise the fingerprint map becomes. During the offline phase, the twice-collected raw CSI fingerprint data accounts for approximately 20% of the total sampling points. Our algorithm is particularly suitable for large-scale indoor dynamic environments, where path planning can significantly reduce labor costs. The objective is to construct fine-grained fingerprint maps with minimal human intervention, distinguishing it from approaches requiring daily fingerprint database updates (Sikeridis et al. 2018). Given the computational resource demands of reinforcement learning, our algorithm improves efficiency by reducing sample size while maintaining practical applicability.

4.4 Generate CSI Fingerprint Algorithm

Table 4.3 Parameter settings of the proposed algorithm

Parameters	Area one	Area two	Area three
Iterations	100	100	100
learning rate	0.01	0.01	0.01
Neurons of input layer	32	32	50
Neurons of hidden layer	64	64	64
Neurons of latent layer	64	32	64
Maximum exploration step length	200	100	300
Number of segments	15	10	10
Number of neighbors	12	5	15

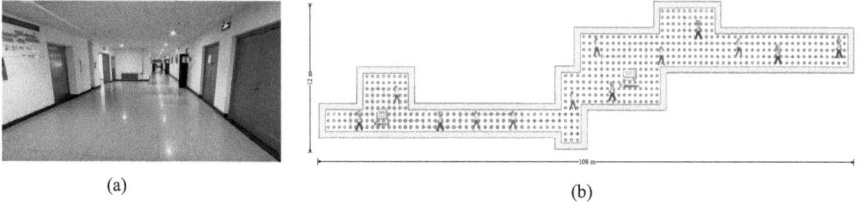

Fig. 4.4 The new experimental scenario. (**a**) The corridor. (**b**) Layout of area three

4.4.7 Performance Evaluation

To evaluate the positioning performance of the proposed algorithm, we conduct a comparative analysis with several representative methods, including Refs. Ciftler et al. (2020), Zhao et al. (2020), Wei et al. (2019), and Zhu et al. (2023). In order to isolate the effect of the fingerprint update process, all fingerprint data in the evaluation are manually collected to ensure consistency in data quality. The average positioning error's CDF is still adopted as the primary evaluation metric, and the parameter settings are listed in Table 4.3.

In addition to the original test areas, a new environment referred to as Area three is introduced to further assess the generalization capability of the algorithm. This area comprises a long indoor corridor covering approximately 108×12 m^2, with 560 sampling points distributed at intervals of 1.2 m, as shown in Fig. 4.4. The data collection process was conducted under realistic conditions, with continuous human movement generating frequent dynamic obstructions. Due to the corridor's linear geometry, the environment exhibits a combination of LoS and NLoS conditions, varying with specific spatial locations. This setting provides a more challenging scenario for positioning algorithms, thus serving as a robust benchmark for evaluating fingerprint update strategies.

(1) Localization Performance Comparisons As shown in Fig. 4.5 and Table 4.4, the proposed scheme consistently outperforms existing state-of-the-art methods across all test environments. In Area one, it achieves a mean localization error of 3.60 m using only 89 sampling points, markedly outperforming the original database

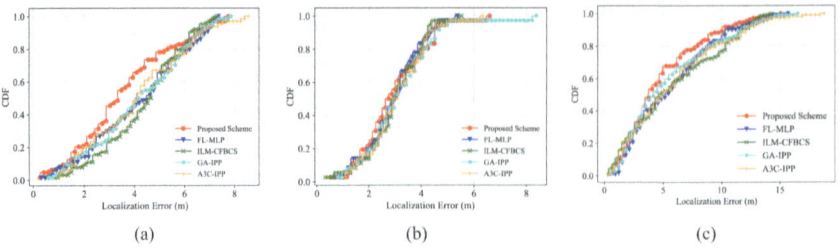

Fig. 4.5 The performance comparison of different algorithms. (**a**) Area one. (**b**) Area two. (**c**) Area three

and other algorithms such as FL-MLP and ILM-CFBCS, which require over three times the data to reach inferior accuracy levels. Notably, ILM-CFBCS, despite its data efficiency, still results in a higher average error of 4.41 m. This highlights the advantage of the proposed method in balancing accuracy and sampling cost. In Area two, where the scene complexity and signal dynamics increase, the proposed method maintains its leading performance, reaching the lowest mean error of 2.98 m with only 47 sampling points. Compared to GA-IPP and A3C-IPP, which utilize over 140 samples, the proposed method not only improves accuracy but also offers a more lightweight and scalable solution for deployment. Figure 4.5b further confirms this advantage through the CDF curves, where the red line representing the proposed scheme shows a sharper rise, indicating a higher probability of achieving low-error predictions. In the most challenging Area three, where propagation conditions and multipath effects are more severe, the proposed approach again demonstrates robust generalization. Despite requiring fewer sampling points than most baselines, it achieves a competitive average error of 4.91 m with lower standard deviation, indicating more stable prediction behavior.

(2) Relation Between Reward and Localization Accuracy On the basis of the previous large-scale generation and sampling strategies, we further evaluate how the agent's path planning affects fingerprint database quality. Leveraging the reward-driven approach introduced in Zhu et al. (2023), we guide data collection using a reward function that accounts for both localization accuracy and the number of exploration steps. To examine this effect, fingerprint databases with different step-length budgets are constructed across multiple areas, and the relationship between total reward and positioning accuracy is analyzed. As shown in Fig. 4.6, higher total rewards typically correspond to better localization performance, with longer step lengths offering more comprehensive data coverage. However, this trend is not strictly linear; beyond a certain point, the accuracy gain diminishes. In Area one, signal blockage leads to the failure of optimal path identification under a 100-step setting, while Area two benefits from clearer signal propagation and achieves better accuracy. In the more complex Area three, increased reward still results in up to 0.8 m improvement in accuracy. Table 4.4 further supports these findings,

4.4 Generate CSI Fingerprint Algorithm

Table 4.4 Performance comparisons with the state-of-the-art researches

Algorithms	Area one			Area two			Area three		
	Sampling points	Mean errors (m)	Std (m)	Sampling points	Mean errors (m)	Std (m)	Sampling points	Mean errors (m)	Std (m)
Proposed scheme	**89**	**3.60**	**1.91**	**47**	**2.98**	**1.26**	**156**	**4.91**	**3.31**
Original database	317	4.68	2.23	176	3.28	1.21	560	4.87	3.49
FL-MLP (Ciftler et al. 2020)	317	4.23	1.92	176	2.95	1.08	560	5.79	3.62
ILM-CFBCS (Zhao et al. 2020)	32	4.41	1.60	18	3.01	1.08	56	5.91	4.02
GA-IPP (Wei et al. 2019)	72	4.21	1.89	57	3.18	1.40	145	5.53	3.89
A3C-IPP (Zhu et al. 2023)	101	4.19	1.95	49	3.12	1.14	174	5.42	4.12

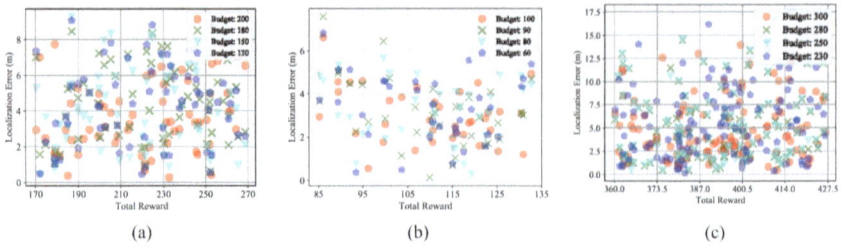

Fig. 4.6 Relation between reward and localization accuracy. (**a**) Area one. (**b**) Area two. (**c**) Area three

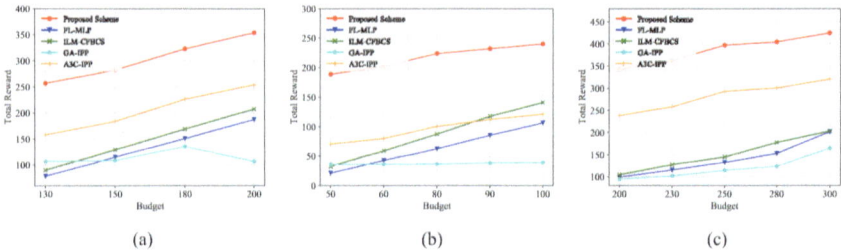

Fig. 4.7 Comparison of rewards with different schemes by budget. (**a**) Area one. (**b**) Area two. (**c**) Area three

confirming the value of reward-guided path sampling in building efficient and accurate fingerprint databases.

(3) Robustness Analysis To evaluate the robustness of different path planning strategies, we analyze the total reward achieved under varying exploration budgets, where the step length is treated as the available sampling capacity. As shown in Fig. 4.7, the proposed method consistently secures the highest total rewards across all three experimental areas, reflecting its ability to adapt to diverse spatial environments and budget constraints. This trend remains stable as the budget increases, indicating that the algorithm can effectively capitalize on additional sampling resources to construct higher-quality fingerprint databases.

Although A3C-IPP performs comparably in some cases, the performance gap becomes more pronounced at larger budgets, where our approach demonstrates superior scalability. In contrast, FL-MLP shows only gradual reward growth, suggesting that its limited exploration ability or CSI update strategy may restrict its effectiveness. ILM-CFBCS performs relatively well in Area two, likely due to reduced signal interference and better spatial openness, but its improvement plateaus at higher budgets. GA-IPP exhibits erratic behavior, with its total rewards fluctuating or even declining in certain settings, which implies susceptibility to local optima. Notably, in Area one, where obstacles and human activity introduce significant environmental noise, our method still outperforms others, indicating strong generalization capability. In Area three, characterized by complex corridor structures, the proposed method efficiently utilizes increased budgets to expand exploration and gather more informative CSI samples, leading to substantial reward gains.

To summarize, this chapter proposes an intelligent CSI fingerprint database update method incorporating imitation learning with minimal human intervention. The approach first collects limited raw data to construct an initial coarse database, then strategically acquires new data samples. The model enhances precision through imitation learning-based data updates, generates the final database via confidence evaluation, and employs KNN for online positioning. Experimental comparisons verify the superiority of the proposed algorithm, including a 73.3% reduction in manual data collection requirements while maintaining high positioning efficiency.

4.5 DBLG Algorithm

4.5.1 Motivation and Challenge

With the rapid development of ISAC technologies in 6G networks, CSI-based fingerprinting has emerged as a key technique for achieving high-precision indoor localization. In practical scenarios, maintaining an accurate and up-to-date fingerprint database during the offline phase is essential, as it directly influences the overall positioning performance. To reduce the reliance on labor-intensive site surveys and improve data quality, recent research has increasingly explored crowdsourcing and prediction-driven approaches for CSI fingerprint updates.

However, crowdsourcing-based fingerprint database updates remain susceptible to noise, incomplete spatial coverage, and environmental dynamics. Maintaining high-accuracy localization requires frequent updates to the fingerprint map, which not only increases overhead but also limits scalability and real-world deployment. Additionally, many existing approaches rely on complex machine learning models for CSI feature extraction, which are often computationally expensive and may generalize poorly when applied to unseen environments, leading to reduced positioning reliability.

To address these challenges, this chapter introduces a novel method for CSI fingerprint updating and prediction, namely DBLG, combining a Deep Broad Learning System (DeepBLS) with a GAN-based data generation module. By integrating confidence coefficients into the learning process, the proposed approach can efficiently generate high-quality CSI fingerprints with improved adaptability and reduced computational cost. The method ensures effective feature extraction from CSI measurements while supporting real-time processing, thereby offering a robust and scalable solution for building reliable fingerprint databases in dynamic indoor environments.

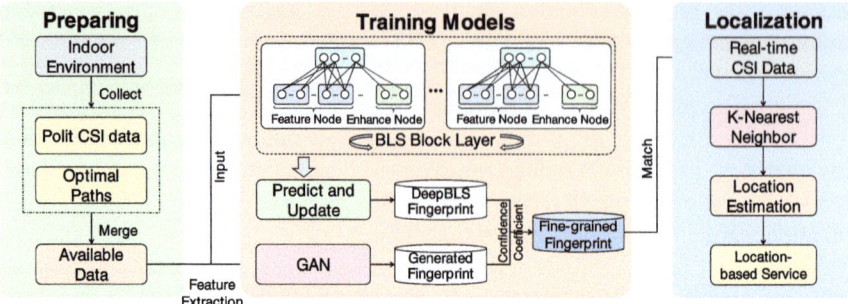

Fig. 4.8 The overall structure of DBLG

4.5.2 The Overview

The overall architecture of the proposed DBLG algorithm is shown in Fig. 4.8. The process begins with the preparation stage, where pilot CSI data and optimized sampling paths are collected from the indoor environment and then merged to form the initial dataset for training. The DeepBLS network is first employed to construct a coarse-grained global CSI fingerprint database. It adopts a layer-wise training strategy, in which the output residual from each layer, adjusted by a learning rate, becomes the input target for the next layer. The training mechanism eliminates the need for backpropagation across layers, thereby reducing computational cost and accelerating the overall learning process.

To further improve fingerprint quality and handle environmental variability, a GAN is incorporated into the framework. It processes the extracted CSI features to perform secondary prediction and refinement, generating high-resolution and realistic fingerprints from limited or noisy samples. It enables the system to adapt effectively to changes in the surrounding environment. The outputs from both DeepBLS and the GAN are evaluated using confidence coefficients, which directly guide the construction of a fine-grained and reliable fingerprint database.

In the online localization stage, the real-time CSI measurements are matched against the refined fingerprint database using the KNN algorithm. The final estimated location is then used to support downstream location-based services. The framework can ensure high-quality fingerprint generation and reliable positioning performance, even under dynamic indoor conditions.

4.5.3 Deep-Broad Learning System Network

The BLS improves feature extraction by increasing the width of the network, enhancing its adaptability to both training and testing data while also boosting generalization capability (Chen and Liu 2017). The architecture consists of three

4.5 DBLG Algorithm

main components: the feature layer, which generates initial feature nodes; the enhancement layer, which applies nonlinear transformations to refine the feature representation; and the output layer, which integrates both types of nodes through a set of connection weights to generate the final output. Importantly, BLS bypasses complex iterative training by only requiring the computation of output weights, which can be efficiently solved as a ridge regression problem.

The DeepBLS adopts a modular structure where each layer functions as an independent BLS block trained with a single forward pass. It setup merges the expressive power of deep architectures with the efficiency of BLS, enabling rapid training and strong generalization. By progressively stacking BLS blocks, the network increases in depth, which enhances both accuracy and robustness across diverse data distributions. The k-th BLS block is initialized using a fixed parameter set ω_0 and trained to generate its respective output. Once training is complete, the output of the first block is denoted by $f_{BLS}(X; \omega_0)$, where X represents the input feature matrix. The residual for training the next block is obtained by subtracting this output from the true labels Y as

$$r^{(k)} = Y - f_{BLS}(X; \omega_0). \tag{4.13}$$

To guide the model toward convergence and prevent oscillation near optimality, each subsequent BLS block receives a scaled version of the preceding residual as its target output. This strategy promotes gradual refinement of predictions. The residual at layer k is calculated as

$$r^{(k)} = Y - f_k(X) = Y - \sum_{i=0}^{k-1} f_{BLS}(X; \omega_0), \tag{4.14}$$

where $f_k(X)$ denotes the cumulative output from the first k BLS blocks. The final model output is the aggregation of all individual block outputs as

$$\begin{aligned} f_k(X) &= \sum_{i=1}^{k-1} f_{BLS}(X; \omega_0) + f_1(X) \\ &= \sum_{i=1}^{k-1} f_{BLS}(X; \omega_0) + \begin{bmatrix} Z^n \mid H^m \end{bmatrix} W^m. \end{aligned} \tag{4.15}$$

The layered training mechanism enables DeepBLS to extract increasingly rich features from raw CSI data. These outputs are subsequently utilized for prediction and updating, supporting the construction of a high-quality global CSI fingerprint database. And Algorithm 1 presents the pseudocode for updating the predicted CSI fingerprint database using DeepBLS.

Algorithm 1 Generate the initial fingerprint database by DeepBLS

1: **Input:** Training data: X_{train}, Y_{train}; Test data: X_{test}; Model parameters: max_iter, learn_rate, NumFea, NumWin, NumEnhan, s, C.
2: **Output:** Final prediction for final_predictions
3: Initialize model parameters and residual mean
4: Create and train the first BLS with X_{train}, Y_{train}
5: Append the trained BLS to model
6: Compute initial prediction f
7: **for** iter $= 1$ to max_iter $- 1$ **do**
8:     Compute residual $r = Y_{train} - f$
9:     Create and train new BLS with X_{train}, $r \times$ learn_rate
10:     Append new BLS to model
11:     Update f with new BLS prediction using Eq. 4.15
12: **end for**
13: Initialize final_predictions as zeros
14: **for** each BLS in model **do**
15:     Predict output for X_{test} using current BLS
16:     Accumulate predictions into final_predictions
17: **end for**
18: **Return** final_predictions

4.5.4 GAN and Confidence Coefficient

The GAN consists of two core components: a generator and a discriminator, as described in Sect. 2.4.4. The generator receives random noise and label information as inputs and produces synthetic data, while the discriminator attempts to distinguish between real and synthetic samples. Through adversarial training, the generator continuously enhances its ability to produce realistic data, whereas the discriminator becomes increasingly adept at identifying subtle differences between real and generated inputs. The overall structure of the GAN is shown in Fig. 4.9.

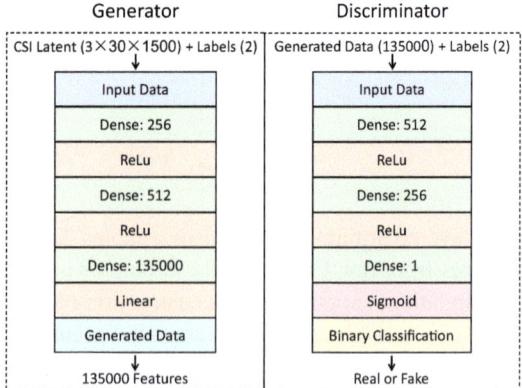

Fig. 4.9 The GAN structure

4.5 DBLG Algorithm

The generator G takes as input a random noise vector z sampled from a Gaussian distribution and a label y, generating synthetic data $\hat{x}$ as

$$\hat{x} = G(z, y), \tag{4.16}$$

where z represents the random latent input, y is the label guiding the generation process, and $\hat{x}$ denotes the resulting synthetic data. The discriminator D evaluates both real data samples x_{real} and generated samples $\hat{x}$, computing the discriminator loss by

$$d_{loss} = \frac{1}{2}\left(D(x_{real}, y) + D(\hat{x}, y)\right). \tag{4.17}$$

The generator loss, which quantifies the ability of G to fool the discriminator, is defined as $g_{loss} = D(G(z, y), y)$. To improve the realism of generated data, a statistical adjustment is applied to align its distribution with that of the real data. Let $\mu_{\hat{x}}$ and $\sigma_{\hat{x}}$ denote the mean and standard deviation of the generated data, and μ_x and σ_x represent those of the real data. The adjusted synthetic data is computed as

$$\text{adjusted generated data} = \left(\frac{\hat{x} - \mu_{\hat{x}}}{\sigma_{\hat{x}}}\right)\sigma_x + \mu_x. \tag{4.18}$$

To ensure high-quality prediction and database refinement, confidence coefficients derived from both the DeepBLS and GAN components are employed during the fingerprint updating process. These coefficients quantify the reliability of each generated fingerprint and are used to fuse the GAN-generated data with DeepBLS predictions:

$$highFinger = c \cdot GAN\_data + (1 - c) \cdot final\_predictions. \tag{4.19}$$

By using the confidence coefficients, the DBLG effectively filters and refines the CSI fingerprint database, resulting in higher accuracy and finer granularity. The refined database better captures subtle variations in wireless channel conditions, thereby improving overall localization performance.

4.5.5 Performance Evaluation

We evaluate the positioning performance of the proposed DBLG algorithm by comparing it with three representative baseline methods (Zhao et al. 2020; Wei et al. 2019; Zhu et al. 2023). To assess the impact of fingerprint updates on localization accuracy, all fingerprint data were manually collected under two distinct scenarios. The CDF of the average positioning error is still used as the primary evaluation metric.

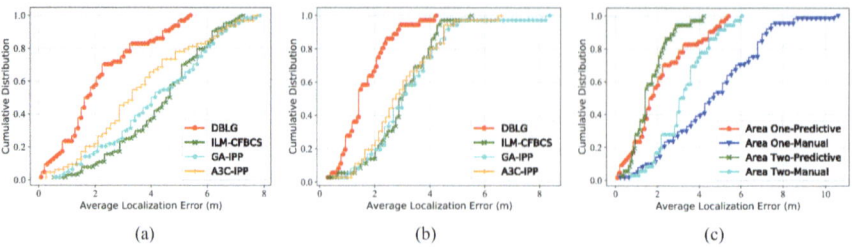

Fig. 4.10 The localization experiments. (**a**) Area one. (**b**) Area two. (**c**) Fingerprint comparison

Table 4.5 Comparisons of localization accuracy and testing time

	Area one			Area two		
Algorithm	Mean error (m)	Std (m)	Time (s)	Mean error (m)	Std (m)	Time (s)
DBLG	**1.99**	**1.42**	**0.01562**	**1.60**	**0.89**	**0.01674**
Manual	4.68	2.22	0.08640	3.28	1.21	0.06903
ILM-CFBCS (Zhao et al. 2020)	4.40	1.59	0.00459	3.01	1.08	0.00481
GA-IPP (Wei et al. 2019)	4.20	1.88	0.00190	3.17	1.40	0.00139
A3C-IPP (Zhu et al. 2023)	4.19	1.95	0.09604	3.11	1.13	0.04951

(1) Localization Comparisons The experimental results verify that the DBLG algorithm achieves superior localization accuracy across both environments, as shown in Fig. 4.10 and summarized in Table 4.5. Compared to the other three algorithms, DBLG yields the lowest average positioning error and standard deviation. Specifically, it improves average localization accuracy by 46.78 and 48.38% in the two scenarios, respectively. Additionally, the lower standard deviation indicates that DBLG exhibits more stable and consistent performance. The improved performance of the updated fingerprint database can be attributed to two main factors. First, the combination of BLS and deep learning enables efficient computation while maintaining accuracy. Second, the model effectively captures discriminative features from the original CSI data, allowing for precise and fine-grained fingerprint updates.

(2) Ablation Study We conduct an ablation study to examine how three core parameters influence the effectiveness of CSI fingerprint database updates and the final localization performance. These parameters include the number of feature nodes NumFea in DeepBLS, the number of training epochs in the GAN module, and the proportion of initial pilot data. Each parameter affects the structure or learning process of the DBLG algorithm in different ways and is therefore analyzed individually. The experimental results are presented in Fig. 4.11.

4.5 DBLG Algorithm

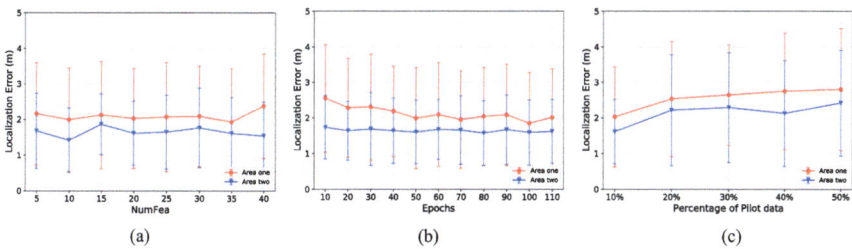

Fig. 4.11 Parameter sensitivity analysis under different configurations. (**a**) Number of features. (**b**) Epoch count. (**c**) Pilot data percentage

First, we evaluate the impact of NumFea, which controls the number of initial features extracted in the BLS model. With other parameters fixed, we vary NumFea from 5 to 40 and observe changes in localization accuracy, as shown in Fig. 4.11a. When it is set to 20, the localization error reaches a stable level in both experimental areas. Specifically, the average error in area one is 2.03 m with a standard deviation of 1.40 m, and in area two the average error is 1.61 m with a standard deviation of 0.90 m. Increasing NumFea beyond the value doesn't further improve accuracy, which confirms that 20 is an effective setting for maintaining model stability and precision.

Second, we assess the effect of GAN training epochs. In this part, the model structure, learning rate, and other parameters remain unchanged while the number of epochs is adjusted from 10 to 110. As illustrated in Fig. 4.11b, the localization error decreases significantly as the number of epochs increases. In area one, the error drops from over 3.5 m to around 2.0 m within 50 epochs. In area two, the error remains consistently low around 1.5 m. When the number of epochs exceeds 50, further improvements in localization accuracy are minimal. It confirms that 50 epochs is sufficient to achieve convergence and satisfactory performance under the current setup.

Finally, we analyze how the proportion of initial pilot data influences the performance of the DBLG algorithm. This pilot data is used to train the initial GPR model for path planning, which guides further CSI data collection. Figure 4.11c shows that increasing the proportion of pilot data from 10 to 50% leads to a gradual increase in localization error. When the proportion is too high, more noise and redundancy are introduced, which degrades the quality of the updated fingerprint database.

To summarize, in this chapter, we propose DBLG, a fine-grained CSI fingerprint database update method that combines DeepBLS and GAN to address the challenge of maintaining accurate fingerprint databases for indoor localization. Building on our previous work (Zhu et al. 2023), we first obtain initial CSI fingerprint data through path planning, and then apply the DBLG algorithm for feature extraction and enhancement. This process produces two global fingerprint databases, which are fused using a confidence-based approach to generate a high-quality CSI fingerprint database. Experimental results show that DBLG outperforms existing methods in both positioning accuracy and efficiency. The proposed method improves localization accuracy by 46.78% and significantly reduces manual effort, making it well-suited for deployment in large-scale indoor scenarios.

References

Chen CP, Liu Z (2017) Broad learning system: an effective and efficient incremental learning system without the need for deep architecture. IEEE Trans Neural Netw Learn Syst 29(1):10–24
Ciftler BS, Albaseer A, Lasla N, Abdallah M (2020) Federated learning for RSS fingerprint-based localization: a privacy-preserving crowdsourcing method. In: International wireless communications and mobile computing, pp 2112–2117
Guo J, Ho IWH, Hou Y, Li Z (2023) FedPos: a federated transfer learning framework for CSI-based Wi-Fi indoor positioning. IEEE Syst J 17(3):4579–4590
Huai S, Liu X, Jiang Y, Dai Y, Wang X, Hu Q (2023) Multi-feature based outdoor fingerprint localization with accuracy enhancement for cellular network. IEEE Trans Instrum Measur 72:5504215
Huan H, Wang K, Xie Y, Zhou L (2022) Indoor location fingerprinting algorithm based on path loss parameter estimation and Bayesian inference. IEEE Sens J 23(3):2507–2521
Jianhua L, Baoshan Z, Songnian L, Zlatanova S, Zhijie Y, Mingchen B, Bing Y, Danqi W (2024) MLA-MFL: a smartphone indoor localization method for fusing multi-source sensors under multiple scene conditions. IEEE Sens J 24:26320–26333
Kipf T, Li Y, Dai H, Zambaldi V, Sanchez-Gonzalez A, Grefenstette E, Kohli P, Battaglia P (2019) Compile: compositional imitation learning and execution. In: International Conference on Machine Learning, pp 3418–3428
Le MT, et al. (2021) Enhanced indoor localization based BLE using Gaussian process regression and improved weighted kNN. IEEE Access 9:143795–143806
Li L, Guo X, Zhang Y, Ansari N, Li H (2022) Long short-term indoor positioning system via evolving knowledge transfer. IEEE Trans Wireless Commun 21(7):5556–5572
Merenda M, Catarinucci L, Colella R, Iero D, Della Corte FG, Carotenuto R (2022) RFID-based indoor positioning using edge machine learning. IEEE J Radio Frequency Identification 6:573–582
Sikeridis D, Rimal BP, Papapanagiotou I, Devetsikiotis M (2018) Unsupervised crowd-assisted learning enabling location-aware facilities. IEEE Internet Things J 5(6):4699–4713
Suroso DJ, Cherntanomwong P, Sooraksa P (2022) Deep generative model-based RSSI synthesis for indoor localization. In: 2022 19th international conference on electrical engineering/electronics, computer, telecommunications and information technology (ECTI-CON), IEEE, New York, pp 1–5
Wang J, Gao Q, Yu Y, Wang H, Jin M (2011) Toward robust indoor localization based on Bayesian filter using chirp-spread-spectrum ranging. IEEE Trans Ind Electron 59(3):1622–1629
Wei Y, Zheng R (2021) Efficient Wi-Fi fingerprint crowdsourcing for indoor localization. IEEE Sens J 22(6):5055–5062
Wei Y, Frincu C, Zheng R (2019) Informative path planning for location fingerprint collection. IEEE Trans Netw Sci Eng 7(3):1633–1644
Wong CC, Feng HM, Kuo KL (2023) Multi-sensor fusion simultaneous localization mapping based on deep reinforcement learning and multi-model adaptive estimation. Sensors 24(1):48
Zhang M, Jia J, Chen J, Deng Y, Wang X, Aghvami AH (2021) Indoor localization fusing Wifi with smartphone inertial sensors using LSTM networks. IEEE Internet Things J 8(17):13608–13623
Zhao M, Qin D, Guo R, Xu G (2020) Research on crowdsourcing network indoor localization based on co-forest and Bayesian compressed sensing. Ad Hoc Netw 105:102176

References

Zheng H, Gao M, Chen Z, Liu XY, Feng X (2019) An adaptive sampling scheme via approximate volume sampling for fingerprint-based indoor localization. IEEE Internet Things J 6(2):2338–2353

Zhu X, Qiu T, Qu W, Zhou X, Wang Y, Wu DO (2023) Path planning for adaptive CSI map construction with A3C in dynamic environments. IEEE Trans Mobile Comput 22(5):2925–2937

Zhu Z, Zhao Y, Tang M, Bu Y, Han H (2025) Robust and effort-efficient image-based indoor localization with generative features. IEEE Trans Mobile Comput 24:6394–6412

Open Access This chapter is licensed under the terms of the Creative Commons Attribution 4.0 International License (http://creativecommons.org/licenses/by/4.0/), which permits use, sharing, adaptation, distribution and reproduction in any medium or format, as long as you give appropriate credit to the original author(s) and the source, provide a link to the Creative Commons license and indicate if changes were made.

The images or other third party material in this chapter are included in the chapter's Creative Commons license, unless indicated otherwise in a credit line to the material. If material is not included in the chapter's Creative Commons license and your intended use is not permitted by statutory regulation or exceeds the permitted use, you will need to obtain permission directly from the copyright holder.

Chapter 5
Accurate Online Data Application

Abstract With the increasing demand for high-precision, low-latency indoor localization in IoT and smart environments, the efficient online processing of CSI data has become a pressing challenge. While Chap. 4 focused on offline fingerprint updates, this chapter turns to online applications, aiming to ensure accurate and rapid localization in dynamic environments. This chapter begins by exploring the fundamental requirements for high-speed localization, emphasizing the trade-offs between speed and accuracy, and the impact of environmental dynamics on real-time performance. We then discuss advanced techniques for accelerating fingerprint matching, including optimized search algorithms and strategies for handling dynamic obstacles. A key highlight is the introduction of the BLS as a lightweight, efficient model for real-time localization, enhanced by ensemble learning to improve robustness and accuracy.

Then, we present the BLS-Location, a lightweight online localization algorithm based on Broad Learning System, which accelerates fingerprint matching and maintains reliable performance with low computational cost. To further enhance adaptability, we propose the ILCL algorithm, which transforms CSI phase data into images and uses a CNN for offline training. During the online stage, it incrementally adapts to new data using a probabilistic method based on BLS, without the need for retraining. Experimental evaluations show that both BLS-Location and ILCL offer significant improvements in accuracy and efficiency, especially in large-scale and dynamic indoor environments.

Keywords Online accurate localization · Dynamic environment adaptation · Broad learning system

5.1 Overview of Real-Time Localization Challenges

5.1.1 Requirements for High-Speed Localization

High-precision real-time localization systems are important in applications requiring rapid and accurate positioning (Lu et al. 2021; Chen et al. 2019b). These systems must complete the entire process from data collection to result output within an extremely short time frame. For example, in intelligent warehousing systems, automated guided vehicles (AGVs) need precise location information within tens of milliseconds to ensure efficient operations. This requirement places significant demands on both the system's processing power and the computational complexity of the algorithms employed.

For WiFi-based localization systems, the system must quickly match new fingerprint data with the existing database (Fernández 2019; Zhu et al. 2020). This process is often limited by computational resources and data processing speed. In large-scale environments, such as shopping malls or airports, traditional matching algorithms struggle to meet real-time demands due to their linear relationship between computational complexity and database size. For instance, when the database contains tens of thousands of fingerprints, traditional methods like the KNN algorithm may take several seconds or longer to complete the matching process. It is insufficient for real-time applications, where rapid response is necessary. The computational burden is particularly pronounced in environments with high-density fingerprint databases. The KNN algorithm requires calculating the distance between the new fingerprint and every fingerprint in the database, which becomes computationally prohibitive as the database size grows. The linear scalability issue is a significant barrier to achieving real-time performance in localization systems.

Moreover, the dynamic nature of indoor environments, where signal propagation can be affected by various factors such as people movement, furniture placement, and multipath effects, further complicates the localization process. These factors introduce variability and noise into the fingerprint data, making it more challenging to achieve accurate and consistent localization results. Therefore, there is a need for more efficient and robust algorithms that can handle large-scale databases and dynamic environments while maintaining high precision.

5.1.2 How to Balance Speed and Accuracy?

In practical localization systems, achieving a balance between positioning speed and accuracy remains a classical challenge. Algorithms with high accuracy often rely on deep learning models or complex optimization processes, which demand considerable computational resources and extended processing time. Such methods may offer superior performance in static evaluations but are often unsuitable for real-time scenarios where response times must be minimized. Therefore, reconciling

5.1 Overview of Real-Time Localization Challenges

Table 5.1 Strategies for balancing speed and accuracy in real-time localization

Strategy	Core idea	Advantages and application scenarios
Lightweight models (e.g., BLS)	Employ efficient structures that avoid iterative training by using random feature mapping and closed-form solutions.	Offers fast training and inference, well-suited for large-scale or frequently updated localization systems.
Model simplification (e.g., PCA, feature selection)	Reduce dimensionality or select discriminative features to decrease computational load.	Enhances inference speed and generalization; useful when input data contains redundancy or noise.
Ensemble learning	Combine multiple base learners to improve accuracy and robustness.	Enables independent training of learners; ideal for dynamic environments requiring stability.
Optimized search algorithms (e.g., KD-Tree Zhou et al. 2008, LSH)	Accelerate fingerprint matching using efficient data structures or hashing.	Decreases search time in large databases; suitable for systems relying on fast nearest-neighbor matching.
Hardware acceleration (e.g., GPU, FPGA)	Leverage parallel architectures to boost processing efficiency.	Significantly improves latency; applicable in time-sensitive scenarios like autonomous robotics or smart manufacturing.

these two competing objectives is essential for the effective real-time localization systems.

To mitigate this contradiction, various strategies have been explored, as given in Table 5.1. One approach is to employ algorithmic optimization techniques that reduce computational overhead while preserving accuracy. Lightweight models, such as BLS (Zhu et al. 2021), offer a promising solution by enabling fast training and inference without sacrificing localization precision. In addition, preprocessing techniques such as dimensionality reduction and feature selection can streamline the input space, resulting in faster execution and more efficient resource utilization. Hardware-level acceleration, using components like GPUs or FPGAs, also plays an important role in improving runtime performance. These methods can be selectively combined according to the requirements of specific deployment scenarios.

Among lightweight models, BLS demonstrates particular advantages in balancing efficiency and effectiveness. Unlike conventional DNNs, which require extensive iterative training, BLS uses a random projection to map inputs into a high-dimensional space, followed by a closed-form solution for output weight computation. This structure eliminates the need for backpropagation, significantly reducing both training duration and computational complexity. Moreover, when integrated with ensemble learning, BLS can further enhance model robustness. By aggregating the outputs of multiple weak learners, ensemble methods achieve higher generalization performance while allowing each learner to be trained independently, thereby reducing time and resource consumption.

5.1.3 Impact of Environmental Dynamics on Real-Time Performance

The dynamic nature of indoor environments poses significant challenges to real-time localization systems. Common changes, such as human movement, rearrangement of furniture, and the introduction of temporary obstacles, continuously alter the wireless signal propagation characteristics. These variations lead to inconsistencies in CSI fingerprint data, resulting in degraded localization accuracy and increased difficulty in maintaining reliable performance under time constraints.

The influence of environmental dynamics on real-time localization performance can be categorized into the following aspects:

1. **Signal Multipath Effects.** In enclosed or cluttered spaces, wireless signals often experience multiple reflections, diffractions, and scattering before reaching the receiver. These multipath effects introduce interference patterns that are difficult to predict and model accurately. As a result, the observed CSI may not correspond directly to the actual position of the device, causing an increase in positioning error. Empirical studies suggest that in typical indoor environments, multipath propagation can raise localization errors by approximately 20% to 30%, especially in scenarios with narrow corridors or high surface reflectivity.
2. **Signal Attenuation and Ambient Noise.** The presence and movement of people, machinery, and mobile objects can introduce sudden changes in signal strength and quality. These changes manifest as attenuation, shadowing, and additional noise in the received signal. Such disruptions reduce the signal-to-noise ratio (SNR), thereby diminishing the reliability of fingerprint-based localization. In experimental setups, it has been observed that attenuation and noise effects may lead to a decline in localization accuracy by 15 to 25%, particularly in high-traffic areas or industrial environments.
3. **Real-Time Database Adaptation.** In dynamic environments, the statistical characteristics of CSI fingerprints evolve over time. To maintain accuracy, the fingerprint database must be updated frequently to reflect the current conditions. This necessity introduces significant pressure on system resources, including memory usage, processing speed, and update latency. Real-time adaptation, without degrading inference speed, becomes a core requirement for maintaining localization performance in continuously changing settings.

To address these challenges, localization systems must adopt adaptive mechanisms capable of responding to environmental changes in real time. One effective strategy is to incorporate incremental or continual learning approaches, which allow the model to integrate newly observed data without undergoing full retraining. These methods update model parameters online, enabling rapid adaptation to fluctuating signal conditions while conserving computational resources. Additionally, real-time data filtering and noise suppression techniques can help mitigate the adverse effects of signal variability. By combining efficient model updating with robust preprocessing, it becomes possible to maintain high localization accuracy and system responsiveness.

5.2 Accelerating Fingerprint Matching Processes

5.2.1 Optimized Search Algorithms for Fast Matching

Traditional algorithms such as KNN, while simple and easy to implement, exhibit linear time complexity with respect to the database size. As the volume of fingerprints increases, the time required for a complete matching process becomes unacceptable for time-sensitive applications. To alleviate this issue, a variety of optimized search algorithms have been developed to enhance processing speed while preserving localization accuracy.

One commonly adopted approach is to build index structures that support efficient similarity queries. Tree-based algorithms, including KD-trees and ball trees, divide the data space hierarchically to reduce the number of comparisons required during matching. These methods are particularly effective in moderate-dimensional spaces and have shown substantial reductions in search time compared to exhaustive search. In scenarios involving high-dimensional data, preliminary dimensionality reduction techniques can be employed to improve the efficiency of tree structures, though this may introduce minor accuracy degradation.

Hashing-based algorithms represent another important category of acceleration techniques. Locality-Sensitive Hashing (LSH), for instance, maps similar fingerprint features to the same or nearby hash buckets, enabling rapid retrieval with sub-linear computational complexity (Zhou et al. 2008). LSH has proven effective in large-scale deployments, such as those involving over 100,000 fingerprints, where it can reduce retrieval latency to a fraction of the time required by traditional KNN methods, while still maintaining localization error within an acceptable range. However, the effectiveness of LSH depends on careful parameter tuning, and it may fail to capture all relevant matches in noisy or ambiguous environments.

Graph-based approximate nearest neighbor methods, such as Hierarchical Navigable Small World (HNSW) graphs, offer another high-performance alternative (Malkov and Yashunin 2018). These algorithms construct a navigable graph structure where each node represents a fingerprint and edges connect closely related entries. Query processing involves traversing the graph starting from an initial point and iteratively refining candidate matches. HNSW-based methods are especially well-suited to dynamic environments, as new fingerprints can be integrated incrementally with minimal computational overhead, supporting efficient updates alongside high-speed search. As indoor localization systems continue to scale up and operate in increasingly dynamic settings, the development and integration of adaptive and robust matching algorithms will remain a key factor for advancing real-time performance.

5.2.2 Handling Dynamic Obstacles and Movements

In dynamic indoor environments, the presence of moving people, opening doors, and mobile obstacles introduces significant variability in wireless signal propagation, which can undermine the reliability and precision of localization systems (Zheng et al. 2019). Such changes often lead to signal occlusion, attenuation, and severe multipath effects, making it challenging for static fingerprint databases to maintain accurate performance over time. These disruptions are particularly evident in scenarios such as airports, shopping malls, or office buildings, where human activity is frequent and unpredictable.

To address these challenges, dynamic adaptation mechanisms are increasingly employed to enable real-time responsiveness. One core strategy involves continuously updating the fingerprint database based on the observed changes in the environment. By incrementally incorporating newly observed data or interpolating between known data points, the system can maintain stability and reduce degradation in accuracy. Interpolation techniques, such as bilinear or kriging-based methods, are often used to estimate CSI fingerprints at unmeasured locations, enhancing spatial coverage without increasing measurement density. These techniques are especially useful when dealing with regions partially occluded by temporary obstacles or where direct measurements are infeasible due to obstruction.

Moreover, in automated fingerprint collection scenarios using robotic platforms, path planning and obstacle avoidance become integral to maintaining efficient and complete data coverage. Robots equipped with wireless sensors and mobility capabilities can autonomously traverse the indoor space and collect CSI fingerprints. To improve navigation and environmental perception, simultaneous localization and mapping (SLAM) algorithms are often integrated. SLAM enables the robot to build an internal map of the environment while simultaneously localizing itself within it. This capability allows the robot to identify and circumvent obstacles in real time, dynamically re-planning its trajectory to ensure thorough and collision-free fingerprint collection, even in environments that are constantly changing (Khairuddin et al. 2015).

In terms of algorithmic support for dynamic adaptation, filtering techniques such as Kalman filtering and particle filtering are widely applied (Simon 2001; Djuric et al. 2003). Kalman filtering, suitable for linear systems with Gaussian noise, can smooth CSI sequences and filter out transient signal fluctuations, thereby stabilizing the localization output. Particle filtering, which is more suitable for nonlinear and non-Gaussian conditions, represents the system's state distribution through a set of weighted samples (particles), allowing for probabilistic tracking of user movement under uncertain and fluctuating signal conditions. Both filtering techniques can be integrated into real-time systems to continuously update localization estimates, reflecting the evolving signal landscape induced by dynamic obstacles.

5.2.3 Light-Weighted Learning Model–Broad Learning System

Broad Learning System (BLS) is a lightweight and efficient machine learning framework well suited for real-time CSI-based localization tasks (Zhu et al. 2021; Chen and Liu 2017). Unlike traditional deep models that require intensive training and a large number of parameters, BLS adopts a flat network structure and leverages ridge regression to directly compute output weights. This architecture not only reduces computational complexity but also accelerates training and inference, making it highly compatible with indoor localization systems that demand low latency and high adaptability.

For the CSI fingerprint learning, BLS begins by constructing a set of feature nodes, as shown in Fig. 5.1. Given a CSI input matrix $H \in \mathbb{R}^{N \times d}$, where N is the number of training samples and d is the feature dimension of each sample, the input is projected into a high-dimensional space using randomly initialized weights and biases. The nonlinear transformation applied to generate the feature nodes is expressed as

$$M_i = \phi(H \cdot W_{ei} + \beta_{ei}), \quad i = 1, 2, \ldots, n, \tag{5.1}$$

where $\phi(\cdot)$ denotes the activation function, typically sigmoid or ReLU, W_{ei} is the weight matrix, β_{ei} is the bias vector, and n is the total number of feature nodes.

Following the generation of feature nodes, BLS introduces enhancement nodes to further enrich the representational capacity of the model. These enhancement nodes are constructed by applying a different nonlinear transformation to the feature node matrix M, using another randomly initialized set of weights and biases. The

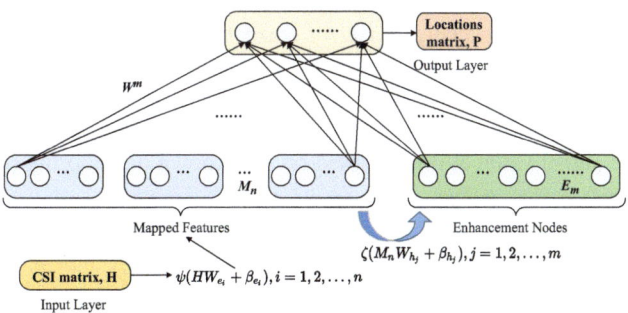

Fig. 5.1 The framework of BLS based on CSI

enhancement nodes are computed as

$$E_j = \tanh(M \cdot W_{hj} + \beta_{hj}), \quad j = 1, 2, \ldots, m, \tag{5.2}$$

where $\tanh(\cdot)$ serves as the activation function, W_{hj} and β_{hj} denote the weights and biases for the enhancement layer, and m is the number of enhancement nodes.

The final feature representation is formed by concatenating the feature nodes M and enhancement nodes E, yielding the matrix $Z = [M \mid E]$, which serves as the input for the output layer. Instead of training the model through iterative gradient descent, BLS calculates the output weights W using closed-form ridge regression as

$$W = (Z^T Z + \lambda I)^{-1} Z^T Y, \tag{5.3}$$

where λ is the regularization parameter, I is the identity matrix, and Y is the target output matrix corresponding to the training samples. The final prediction result is then obtained as $\hat{Y} = Z \cdot W$.

It enables BLS to achieve efficient and accurate learning from CSI measurements with minimal computational overhead. Furthermore, BLS supports incremental learning, allowing the system to adapt to newly collected data without retraining the entire model. This is beneficial in dynamic indoor environments where wireless conditions may change frequently due to user movement or structural alterations. In addition, BLS not only enables fast fingerprint matching but also enhances localization accuracy, which will be further discussed in Sect. 5.4.

5.3 Techniques for Enhancing Localization Accuracy

5.3.1 Error Correction and Refinement

Localization accuracy can be significantly affected by signal noise, multipath effects, and sudden environmental changes. To mitigate these factors, researchers have proposed various strategies for error correction and refinement, with a focus on enhancing CSI data quality and improving location estimation.

One common approach is interpolation, which compensates for missing or irregular CSI samples. A simple yet effective method is linear interpolation, which estimates the signal value at time t based on two known neighboring points t_1 and t_2 by

$$\hat{H}(t) = H(t_1) + \frac{H(t_2) - H(t_1)}{t_2 - t_1}(t - t_1). \tag{5.4}$$

It can smooth sudden signal drops and ensure continuity in fingerprint data. More advanced interpolation methods, such as spline or polynomial interpolation, can be used to preserve higher-order trends in the data.

Filtering methods are also essential for noise suppression. Kalman filtering is a widely used linear estimator that recursively predicts and updates the state of a system based on measurements. The predicted state $\hat{x}_{k|k-1}$ is computed by projecting the previous estimate using a state transition model as

$$\hat{x}_{k|k-1} = F_k \hat{x}_{k-1|k-1} + B_k u_k, \tag{5.5}$$

where F_k is the state transition matrix, B_k is the control input model, and u_k is the control input. After prediction, the algorithm evaluates the uncertainty of this estimate with a covariance matrix update. Once a new measurement z_k is obtained, the Kalman gain is calculated to determine how much the measurement should influence the updated state by

$$K_k = P_{k|k-1} H_k^T (H_k P_{k|k-1} H_k^T + R_k)^{-1}. \tag{5.6}$$

This gain adjusts the predicted state based on the difference between the measurement and the predicted observation. The updated state estimate is then refined as $\hat{x}_{k|k} = \hat{x}_{k|k-1} + K_k(z_k - H_k \hat{x}_{k|k-1})$. Kalman filtering is effective in relatively linear systems.

However, for environments with strong non-linearities or unpredictable dynamics, particle filtering may offer better performance. This approach uses a set of weighted particles to represent the probability distribution of the system state. The importance weight of each particle $x_k^{(i)}$ is evaluated based on the likelihood of the observation as

$$w_k^{(i)} \propto p(z_k | x_k^{(i)}). \tag{5.7}$$

After normalization, the final estimate is given as a weighted average $\hat{x}_k = \sum_{i=1}^{N} w_k^{(i)} x_k^{(i)}$. It is robust in the presence of signal fluctuations and non-Gaussian noise, which are common in indoor wireless environments. In addition to temporal correction, spatial constraints can also be introduced. For instance, map-based pruning can eliminate physically implausible locations from consideration, and probabilistic fusion with inertial sensors or environmental priors can further refine predictions.

5.3.2 Ensemble Learning for Robust Fingerprint Matching

Ensemble learning is a machine learning strategy that improves model robustness and predictive performance by aggregating multiple individual learners (Sagi and

Rokach 2018). Instead of relying on a single model, ensemble methods integrate the outputs of diverse models to reduce variance, correct biases, and improve generalization. It's beneficial for indoor localization systems based on CSI, where signal characteristics are often unstable due to multipath effects, human movement, and environmental dynamics (Taniuchi and Maekawa 2014).

A representative example of this approach is the EnsemLoca algorithm proposed by Wu et al. (2021), which combines a BLS-based ensemble learning framework. The multiple base learners are trained independently using bootstrapped subsets of the original training data. Bootstrapping involves sampling with replacement to generate diverse training sets, encouraging the base models to learn complementary patterns. Each learner is constructed using the BLS, known for its fast training and low computational complexity. And it ensures that the ensemble remains lightweight while achieving substantial performance gains.

The training process begins with data preprocessing, where raw CSI measurements are normalized to suppress noise and outliers. It improves the stability of subsequent learning. Then, a set of base learners is built. For each learner, a subset of the input–label pairs is sampled to form training data (C_k, Y_k). Each learner processes its respective subset using the BLS framework, and its output is denoted as

$$F_k = BLS(C_k, Y_k), \quad k = 1, 2, \ldots, n, \tag{5.8}$$

where F_k represents the feature representation or prediction from the k-th base model. All base learners are trained in parallel, using the BLS model's efficiency. Since BLS avoids iterative gradient-based optimization, the overall training time remains low even when the ensemble size increases. After obtaining the outputs $F_1, F_2, \ldots, F_n$, they are concatenated and used as the input to a meta-learner or combiner. The combiner is typically another BLS model trained to map the aggregated base outputs to the final prediction. Its weight matrix W_{stack} is computed through

$$W_{stack} = BLS([F_1, F_2, \ldots, F_n], Y), \tag{5.9}$$

where Y denotes the true labels corresponding to the original training data. The meta-learning allows the system to adaptively weigh the contributions of different base learners based on their relevance to the target outputs.

During inference, the ensemble system proceeds in two stages. First, each trained base model generates an output from the test input. These outputs are then passed to the combiner, which produces the final localization prediction as

$$\hat{Y} = F_{stack}(F_1, F_2, \ldots, F_n, W_{stack}). \tag{5.10}$$

It improves robustness against environmental changes and unseen signal variations, particularly in NLoS scenario. Experimental results have shown that the EnsemLoca system achieves notable gains in localization performance. Specifically, it improves

5.3 Techniques for Enhancing Localization Accuracy

accuracy by 13 and 23% in LoS and NLoS scenarios, respectively, compared to a single BLS model.

5.3.3 How to Handle the New Fingerprint Data?

Effectively managing new fingerprint data is important for maintaining the performance and adaptability of localization systems, especially in complex indoor environments that frequently undergo structural or contextual changes. To address this need, several strategies have been developed, each offering distinct mechanisms, benefits, and limitations. These strategies can be broadly categorized into incremental learning, data augmentation, active learning, and transfer learning.

(1) Incremental Learning allows the model to incorporate new fingerprint data through continuous updates of its parameters (Van de Ven et al. 2022). Unlike traditional methods that require retraining with the entire dataset, this approach leverages online algorithms such as stochastic gradient descent to modify model weights based on newly observed samples. The update process typically follows a gradient-based optimization:

$$W_{\text{new}} = W_{\text{old}} - \eta \nabla L(W_{\text{old}}, X_{\text{new}}, Y_{\text{new}}), \tag{5.11}$$

where η denotes the learning rate, and $(X_{\text{new}}, Y_{\text{new}})$ represents the new data point. This design supports real-time model adaptation, which is especially useful in environments with frequent layout changes or human activity. To prevent degradation of previously learned knowledge, techniques such as regularization and memory replay are often applied.

(2) Data Augmentation enhances the diversity and representativeness of training samples by introducing synthetic variations. Common strategies include adding Gaussian noise to emulate motion interference, simulating multipath effects, or altering signal amplitude and temporal patterns. These modifications expose the model to a wider range of scenarios, improving its generalization ability and robustness. It is beneficial during system initialization or in scenarios with insufficient real-world data. Nevertheless, care must be taken to ensure that the augmented samples do not introduce unrealistic or misleading signal patterns.

(3) Active Learning improves labeling efficiency by selectively querying the most informative samples (Ren et al. 2021). Rather than passively accepting randomly collected data, this method evaluates prediction uncertainty or data diversity to prioritize samples that are likely to refine decision boundaries. By focusing annotation efforts on ambiguous or underrepresented regions of the feature space, active learning accelerates model improvement while minimizing labeling costs. It is well-suited for complex or large-scale indoor environments where comprehensive data collection is infeasible. However, effective implementation depends on accurate uncertainty estimation and the availability of a sufficiently capable initial model.

(4) Transfer Learning facilitates system deployment in new environments by reusing knowledge learned from a different but related context (Niu et al. 2021). A pre-trained model is partially or fully adapted to the new domain using a smaller dataset, often through fine-tuning specific network layers while preserving general representations. This approach greatly reduces the need for extensive data collection and training time. It is commonly applied in large-scale or multi-site localization systems where environmental similarities can be leveraged. Challenges may arise when the source and target environments differ significantly, which can lead to performance degradation if the transferred knowledge does not align with the new signal characteristics.

To summarize, a comparative overview of these methods is provided in Table 5.2, which summarizes their core ideas, advantages, typical application contexts, and potential challenges. Each method offers unique benefits and can be selected or combined based on the characteristics and demands of the deployment environment.

5.4 BLS-Location Algorithm

5.4.1 Motivation and Challenge

With the increasing demand for high-precision indoor localization in 6G ISAC, CSI-based fingerprinting has gained substantial attention due to its fine-grained representation of wireless environments (Zhang et al. 2023; Umer et al. 2025). However, the construction and maintenance of a high-quality fingerprint database remain challenging. In practical deployments, raw CSI data often suffer from missing entries and measurement noise, which degrade the quality of the fingerprint map. These imperfections arise from various factors such as unstable hardware responses, environmental interference, and user-induced variability during the data collection process. As a result, the accuracy and reliability of existing localization systems are significantly limited, particularly in dynamic or densely populated indoor environments.

In addition to data quality issues, most CSI fingerprinting algorithms face considerable computational burdens during the offline training phase. Many state-of-the-art approaches rely on deep learning or iterative optimization schemes to model the complex relationship between CSI measurements and physical locations. While such methods can achieve high accuracy in ideal settings, they typically involve time-consuming training processes and require substantial computational resources. This severely restricts their scalability and hinders real-time adaptability when the environment changes. Moreover, the lack of efficient preprocessing and generalizable learning frameworks has further impeded the development of lightweight and robust localization systems. Despite growing research efforts, a unified and practical solution to these challenges remains elusive.

5.4 BLS-Location Algorithm

Table 5.2 Comparison of strategies for handling new fingerprint data

Strategy	Core idea	Advantages	Application scenarios	Challenges
Incremental learning	Continuously update model parameters as new fingerprint data becomes available, often using online learning algorithms like SGD.	Facilitates real-time adaptation to environmental changes and reduces retraining costs.	Suitable for dynamic environments such as offices or warehouses with frequent layout or occupancy changes.	Risk of model drift or catastrophic forgetting; requires careful design of update mechanisms and stability preservation.
Data augmentation	Enhance the training dataset by applying signal transformations such as noise injection, scaling, or synthetic multipath modeling.	Improves model robustness and reduces overfitting to specific environmental conditions.	Ideal for early-stage systems with limited real-world data or for enhancing diversity in synthetic datasets.	May generate unrealistic samples if transformation parameters are poorly chosen; risk of introducing noise patterns not present in the target domain.
Active learning	Select the most informative unlabeled samples for annotation based on uncertainty or diversity metrics.	Maximizes learning efficiency while minimizing the need for manual labeling.	Effective in large-scale or complex sites (e.g., hospitals, stations) where full labeling is expensive or impractical.	Requires a reliable uncertainty estimation method; initial model performance must be sufficient to guide selection.
Transfer learning	Leverage knowledge from a model trained on a source domain by fine-tuning it with a small amount of target domain data.	Reduces data and training demands; accelerates deployment in similar environments.	Useful for scaling systems to multiple sites with structural similarities, such as university buildings or retail chains.	Negative transfer can occur if source and target domains differ significantly; requires careful selection of transferable layers.

To address these issues, we propose a novel CSI fingerprinting method named BLS-Location, which utilizes a broad learning system to construct and update the localization model efficiently (Zhu et al. 2021). The proposed approach integrates a uniform preprocessing pipeline based on Kalman filtering, expectation-maximization (EM), and principal component analysis to handle missing and noisy data. In the offline phase, the BLS framework enables fast weight computation without iterative training, greatly reducing the time required to build the model. Moreover, we introduce a kernel position mechanism that employs a naive Bayes classifier to enhance the final location estimation. Experimental results in two typical indoor scenarios show that BLS-Location achieves high localization accuracy with low computational complexity, outperforming several baseline methods in both robustness and efficiency.

5.4.2 The Overview

The overall architecture of BLS-Location is illustrated in Fig. 5.2, which consists of a preprocessing module, an offline training stage based on the BLS, and an online localization phase integrating prediction and probabilistic refinement. The process begins with a transmitter–receiver pair collecting 50 packets of CSI data to form the original CSI matrix. Given the common issues of noise and missing values in real-world measurements, a dedicated preprocessing module is designed to enhance data quality. Specifically, PCA is first applied to reconstruct the data structure, followed by a Kalman filter to eliminate noise. To further compensate for incomplete observations and reinforce temporal consistency, the EM algorithm and a smoothed filter are employed, ultimately producing a refined CSI matrix suitable for model training.

In the offline phase, the reconstructed CSI matrix serves as the input to the BLS, which efficiently learns the mapping between signal features and physical

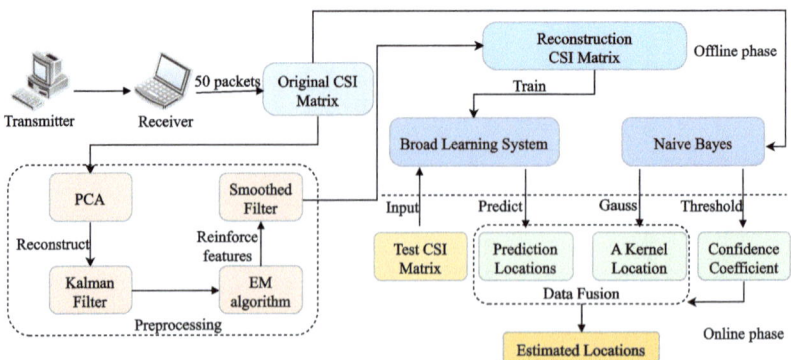

Fig. 5.2 BLS-location architecture

5.4 BLS-Location Algorithm

locations. Owing to the flat structure of BLS, the training process avoids the iterative backpropagation typical of deep neural networks, thereby reducing computational overhead. Additionally, a naive Bayes model is trained in parallel to establish a probabilistic representation of location distributions, which is later used to assign confidence coefficients during the online phase.

In the online phase, the test CSI matrix is fed into the trained BLS to generate initial prediction locations. These predictions are then fused with the kernel location derived from the naive Bayes model using a Gaussian-based strategy. The final location output is determined by incorporating a confidence coefficient, ensuring robust decision-making under uncertain or noisy conditions.

5.4.3 CSI Data Preprocessing

In this chapter, PCA is used to reconstruct the original CSI matrix, denoted as H', rather than for dimensionality reduction. This approach helps suppress noise while retaining key structural features. Considering that certain types of noise may implicitly reflect environmental characteristics, we retain them through reconstruction. The PCA reconstruction is defined as

$$\text{PCA}(H, \alpha) \rightarrow H', \tag{5.12}$$

where H denotes the original CSI matrix, and $\alpha = 95\%$ represents the proportion of variance retained, ensuring most of the signal information is preserved.

To further mitigate the impact of noise and data loss, the Kalman filter is combined with the EM algorithm using the `pykalman` library (Daniel 2012). The EM algorithm iteratively estimates missing values through maximum likelihood, while the Kalman filter predicts the current state based on prior observations, effectively filtering the noise. Unlike methods that rely on strong statistical assumptions about data distributions, our strategy directly minimizes prediction errors. The output of the filtering process is

$$\text{filter}(H') \rightarrow Filter\_mean, Filter\_cov, \tag{5.13}$$

where $Filter\_mean$ is the filtered CSI data and $Filter\_cov$ denotes the covariance matrix. A smoothing step is finally applied to ensure the results are consistent with actual signal propagation dynamics. And the pseudocode of preprocessing is given in Algorithm 2.

Algorithm 2 Data preprocessing of BLS-location

1: **Input:** 50 packets of raw CSI matrix H with the size of 3×30 for each of the N locations; PCA percentage α; transition matrix A; observation matrix B; iteration number iterNum
2: **Output:** Reconstruction CSI matrix X
3: Set $A = [1]$, $B = [1]$, iterNum $= 10$
4: Construct a smoothed filter using Butterworth
5: **for** $i = 1$ to N **do**
6:     Apply PCA to H_i: (lowDim$_i$, H'_i) = PCA(H_i, α)   ▷ H'_i is a matrix with the size of $3 \times 30 \times 50$
7:     Apply element-wise absolute value: H'_i = abs(H'_i) ▷ abs($\cdot$) is the absolute value function
8:     Initialize Kalman Filter: kf = KalmanFilter(A, B)
9:     **for** $j = 1$ to iterNum **do**
10:         Update Kalman Filter with EM: $kf = kf.\text{em}(H'_i)$ ▷ EM algorithm processing
11:     **end for**
12:     Apply Kalman filtering: (Filter_mean$_i$, Filter_cov$_i$) = $kf.\text{filter}(H'_i)$
13:     Compute X_i using Butterworth filter with input Filter_mean$_i$
14: **end for**
15: **Return** X_i ▷ X_i is the matrix with N rows and 4500 columns

5.4.4 The Training Model with BLS and A Probabilistic Method

The training phase of the BLS-Location algorithm involves both BLS model training and a probabilistic method based on naive Bayes, as shown in Fig. 5.1. The process begins by normalizing the preprocessed CSI data using

$$Z = \frac{X - \mu}{\sigma}, \qquad (5.14)$$

where X is the original data, and μ and σ represent the mean and standard deviation, respectively. This normalization ensures that the data is appropriately scaled for subsequent processing. The normalized data is then transformed into feature nodes through the tanh activation function as

$$M_i = \psi_i(H \cdot W_{e_i} + \beta_{e_i}), \quad i = 1, 2, \ldots, n, \qquad (5.15)$$

where W_{e_i} and β_{e_i} are randomly generated weights and biases. These feature nodes capture the essential characteristics of the input data.

To enhance the model's representational power, enhancement nodes are generated via nonlinear transformations by

$$E_j = \zeta_j(M^n \cdot W_{h_j} + \beta_{h_j}), \quad j = 1, 2, \ldots, m, \qquad (5.16)$$

where ζ_j is also the tanh function. These enhancement nodes provide additional features that improve the model's accuracy. The feature nodes and enhancement nodes are then connected to the output layer, and the weights W_m are computed

5.4 BLS-Location Algorithm

Algorithm 3 BLS-location trains the weights

1: **Input:** N groups of the reconstructed CSI matrix X and the raw CSI matrix H
2: **Output:** weight matrix W and a kernel position $\hat{L}$
3: Normalize X according to Eq. 5.14
4: Split samples for training with n groups by cross-validation
5: **for** $i = 1$ to n **do**
6:     **for** $k = 1$ to epoch **do**
7:         Randomly initialize W_{e_i} and β_{e_i}
8:         $M_i = \psi_i(X_i \times W_{e_i} + \beta_{e_i})$ ▷ Based on the tanh activation function
9:         **for** $j = 1$ to m **do**
10:             Randomly initialize W_{h_j} and β_{h_j}
11:             $E_j = \zeta_j(M_i \cdot W_{h_j} + \beta_{h_j})$ ▷ Based on the tanh activation function
12:             Compute $Y_i = [M_i | E_j] \cdot W_j$ ▷ Y is the matrix with N rows and 2 columns
13:             $W_j = [M_i | E_j]^+ \cdot Y_i$
14:         **end for**
15:     **end for**
16: **end for**
17: **Obtain the kernel position**
18: Compute the Gaussian distribution $Gauss$ of each sample for all locations
19: Compute the variance σ of the raw CSI matrix H
20: **for** $i = 1$ to N **do**
21:     $\hat{H}_i = 1/(1 + \exp(-H_i \cdot Gauss_i))$ ▷ $\hat{H}_i$ is the matrix with N rows and 4500 columns
22:     $p_i = \exp(-\frac{1}{\lambda \sigma}\sqrt{(H_i - \hat{H}_i)^2})$ ▷ Compute the posterior probability
23:     $\hat{P}_i = p_i \Big/ \sum_{i=1}^{N} p_i$
24: **end for**
25: **Return** The kernel position $\hat{L}$

---

using the pseudo-inverse $Y = [M_n | E_1, E_2, \ldots, E_m] \cdot W_m = [M^n | E^m] \cdot W^m$, producing the final output used for localization.

To further improve localization performance, a naive Bayes approach is employed to compute the kernel location $\hat{L}$. The posterior probability of each location L_i is calculated as

$$P(L_i | X) = \frac{P(X | L_i) \cdot P(L_i)}{P(X)}, \quad (5.17)$$

where $P(L_i | X)$ is the probability of observing CSI data X at location L_i, and $P(L_i)$ is the prior probability of L_i. The kernel location is then computed as

$$\hat{L} = \sum_{i=1}^{N} \frac{P(L_i | X)}{\text{sum}(P(L_i | X))} \cdot L_i. \quad (5.18)$$

By combining the strengths of BLS and naive Bayes, the BLS-Location algorithm improves localization accuracy. And the corresponding pseudocode is presented in Algorithm 3.

5.4.5 Online Localization Algorithm

Once the training of the BLS-Location model is complete, location estimates can be obtained using the test set. Unlike traditional methods, the BLS-Location algorithm integrates machine learning-based regression and a probabilistic framework during the online phase as part of a data fusion strategy. To make the final decision on the location estimation, we introduce a confidence coefficient, which is calculated as

$$c_i = \frac{Gauss_i}{\sum_{i=1}^{N} Gauss_i}, \tag{5.19}$$

where $Gauss_i$ is the Gaussian value for each sample, calculated using the Gauss function applied to the raw CSI matrix. This confidence coefficient c_i is conceptually similar to the posterior probability P_i, but it differs in that the noise propagation is non-uniform across space. The impact of this noise on accuracy varies spatially.

To refine the confidence coefficient, a thresholding mechanism is applied. This is done by comparing c_i with the mean confidence coefficient, $mean$, and adjusting it based on the following logic by

$$\hat{c}_i = \begin{cases} threshold & \text{if } c_i \leq mean \\ 1 - threshold & \text{if } c_i > mean \end{cases}, \tag{5.20}$$

where $mean$ represents the average value of all confidence coefficients, and $threshold \in [0, 1]$ is a tunable parameter that helps to balance the confidence level.

To refine the localization estimate, the final estimated location L_i is obtained by a weighted fusion of the regression result Y_i and the kernel location $\hat{L}$:

$$L_i = \hat{c}_i \times \hat{L} + (1 - \hat{c}_i) \times Y_i, \tag{5.21}$$

where $\hat{c}_i$ is the normalized confidence coefficient. This fusion strategy leverages the strengths of both the regression result and the probabilistic kernel location, providing a more accurate and robust estimate of the location.

5.4.6 Performance Evaluation

We still use the previous experimental scenarios to verify the performance of the proposed BLS-Location, and compare it with the state-of-the-art localization methods including HORUS (Youssef and Agrawala 2005), RADAR (Bahl et al. 2000), SWIM (Chen et al. 2019a), and BAYES (Fernández 2019).

(1) Localization Comparisons The experimental results, as shown in Table 5.3 and Fig. 5.3, show that BLS-Location achieves the best overall performance in both

5.4 BLS-Location Algorithm

Table 5.3 Localization accuracy, training time and testing time

Algorithms	Area one				Area two			
	Training time (s)	Testing time (s)	Mean errors (m)	Std (m)	Training time (s)	Testing time (s)	Mean errors (m)	Std (m)
BLS-location	**0.88**	0.004	**3.47**	1.73	**0.23**	0.002	**2.39**	1.27
MLP	4.58	0.003	4.62	2.06	5.67	0.003	2.71	1.37
NN	47.68	0.021	4.06	**1.70**	21.95	0.020	2.97	1.27
LSTM	7706.12	0.047	7.48	3.68	4904.17	0.043	5.54	2.18
MOR	40.88	0.014	3.72	1.70	23.10	0.009	2.43	**1.25**
HORUS (Youssef and Agrawala 2005)	N/A	0.002	3.90	1.84	N/A	**0.001**	2.62	1.42
RADAR (Bahl et al. 2000)	N/A	0.070	3.81	2.08	N/A	0.036	3.07	1.37
SWIM (Chen et al. 2019a)	0.99	0.088	4.58	2.40	0.48	0.032	3.47	2.17
BAYES (Fernández 2019)	3.87	**0.001**	4.56	2.17	3.68	0.001	2.87	1.46

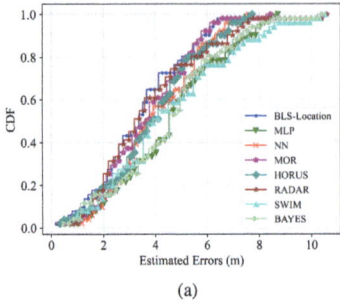 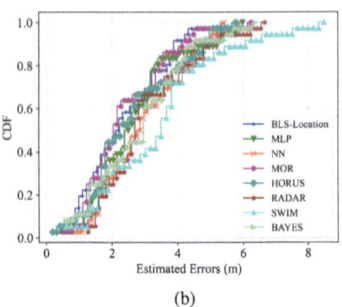

Fig. 5.3 The localization performance comparison. (**a**) Area one. (**b**) Area two

indoor environments. In Area one, it reaches a mean localization error of 3.47 m with a standard deviation of 1.73 m. In Area two, its mean error is further reduced to 2.39 m with a standard deviation of 1.27 m. These results indicate that BLS-Location not only provides accurate positioning but also maintains good stability across environments. Compared to other learning-based methods, BLS-Location improves significantly on both accuracy and efficiency. The MLP model records errors of 4.62 m in Area one and 2.71 m in Area two. The NN algorithm performs slightly better in Area one at 4.06 m but is less stable. Although MOR achieves similar standard deviations, its accuracy is lower than that of BLS-Location in both settings, with 3.72 m in Area one and 2.43 m in Area two. LSTM, on the other hand, shows poor results in both accuracy and stability, with errors of 7.48 and 5.54 m, reflecting the challenges of training deep sequential models on limited and noisy fingerprint data.

For the training efficiency, BLS-Location clearly outperforms the alternatives. It completes training in 0.88 seconds in Area one and 0.23 seconds in Area two. It's significantly faster than MOR, which requires over 40 seconds, and orders of magnitude faster than LSTM, which takes more than 7700 seconds in Area one and over 4900 seconds in Area two. SWIM is relatively efficient, with training times around 0.99 and 0.48 seconds, but its localization accuracy is considerably worse, especially in Area two where the mean error reaches 3.47 m and the standard deviation exceeds 2.16 m. Regarding testing time, all algorithms perform adequately for real-time applications, but some variations are observed. BLS-Location maintains a low testing time of 0.004 seconds in Area one and 0.002 seconds in Area two. Although BAYES and HORUS achieve slightly lower values, their overall accuracy is inferior. For instance, HORUS produces an error of 3.90 m in Area one and 2.62 m in Area two. RADAR and SWIM both suffer from higher variability, and LSTM again falls short due to its long inference time and low precision.

(2) Ablation Experiments To evaluate the impact of the kernel position, we compare BLS-Location with a baseline version that uses BLS without the kernel position module. As shown in Fig. 5.4a, BLS-Location achieves better localization

5.4 BLS-Location Algorithm

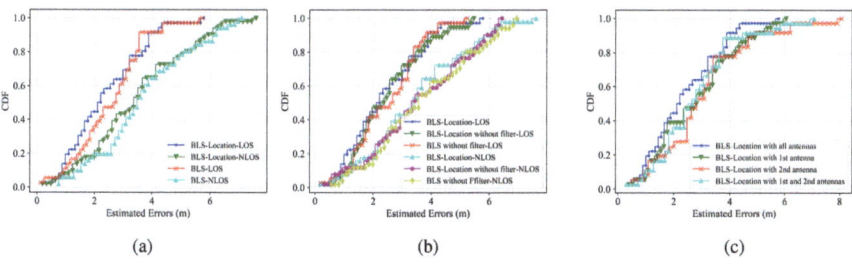

Fig. 5.4 The performance comparison of different settings. (**a**) Kernel position. (**b**) Preprocessing. (**c**) Different antennas

accuracy than BLS in both Area one and Area two. In Area one, the mean error decreases from 3.70 to 3.47 m, and in Area two, it drops from 2.55 to 2.39 m. The improvement is more noticeable in the more complex environment of Area one, where obstacles lead to greater signal variation. In contrast, Area two benefits from fewer obstructions, resulting in generally lower localization errors. The kernel position helps enhance feature representation by refining the regression process, leading to more accurate predictions. Although the standard deviation of errors for BLS is slightly smaller, the difference is marginal. Inference time remains similar between the two schemes, which verifies that the improved accuracy of BLS-Location does not require additional computation.

Furthermore, we compare the performance of BLS-Location with and without preprocessing in both experimental areas, and Fig. 5.4b gives the results. In Area two, about 40% of the test samples achieve mean errors below 2 m across all schemes, while in Area one, only 20% fall within this threshold, indicating the greater complexity of that environment. BLS-Location consistently shows better accuracy than other schemes, even when preprocessing is not applied. Nevertheless, the differences among methods without preprocessing are relatively small, and using preprocessing clearly improves accuracy, especially in Area one where NLoS conditions are more prominent.

Finally, We evaluate the impact of antenna quantity on the performance of BLS-Location by comparing three configurations in Area two, as shown in Fig. 5.4c. The first uses 30 CSI values from a single antenna, the second uses 60 CSI values from the first and second antennas, and the third includes 90 CSI values from all three antennas. Results show that incorporating all three antennas leads to the best performance, with a mean localization error of 2.39 m. About 60% of test samples achieve errors under 2.5 m in this setting, while this proportion drops below 50% when using only the first antenna. The two-antenna setup also performs worse, mainly due to the limited contribution of the second antenna. Although expanding the input from 30 to 90 CSI values increases data size, the training time remains efficient, requiring less than one second even with three antennas. Without preprocessing, the error for the three-antenna configuration is 2.50 m, and preprocessing improves accuracy by approximately 10 cm. The single and dual antenna configurations make larger errors, with maximum deviations reaching around 0.62 m.

To summarize, in this chapter, we proposed BLS-Location, a fast and accurate wireless fingerprinting algorithm for indoor localization based on CSI data. The proposed approach incorporates a unified preprocessing strategy that reconstructs the CSI using PCA without dimensionality reduction, followed by noise smoothing with the Kalman filter and the EM algorithm. In the offline phase, BLS is employed to extract essential features from the CSI fingerprints and to train the localization model. A probabilistic method based on naive Bayes is further introduced to estimate the kernel position, which enhances prediction accuracy during the online phase. And the experimental results show that BLS-Location achieves high localization accuracy and fast inference speed compared to several existing methods.

5.5 ILCL Algorithm

5.5.1 *Preliminaries*

With the continued expansion of location-based services and the increasing importance of precise indoor localization in 6G ISAC, fingerprint-based methods have emerged as a prominent solution due to their cost-effectiveness and wide accessibility (Roy and Chowdhury 2021; Umer et al. 2025). Among them, CSI-based localization schemes stand out for their ability to capture fine-grained wireless signal characteristics. However, the existing methods have two prominent challenges. First, maintaining the accuracy of the fingerprint database becomes difficult as the environment evolves or as new data are introduced. Changes such as user movement, layout alterations, or time-varying signal behaviors can significantly degrade model performance unless the system is frequently retrained. This retraining process is not only computationally intensive but also time-consuming, limiting the scalability and adaptability of existing approaches.

Second, while many localization methods adopt deep learning frameworks to extract high-level representations from CSI measurements, these models typically require complete retraining when new samples are added. This inflexibility creates a bottleneck in dynamic environments, where rapid updates to the fingerprint database are necessary to preserve localization performance. Moreover, retraining introduces latency and consumes significant computing resources, which are often impractical in large-scale or resource-constrained deployments.

To overcome these limitations, we propose a novel indoor localization framework namely ILCL, which combines the CNN for spatial feature extraction with the efficiency of BLS for fast model updates (Zhu et al. 2022). In the offline phase, CSI phase information is captured through a customized driver and transformed into image-like representations suitable for CNN-based feature learning. Instead of retraining the entire model from scratch, ILCL utilizes an incremental learning

5.5 ILCL Algorithm

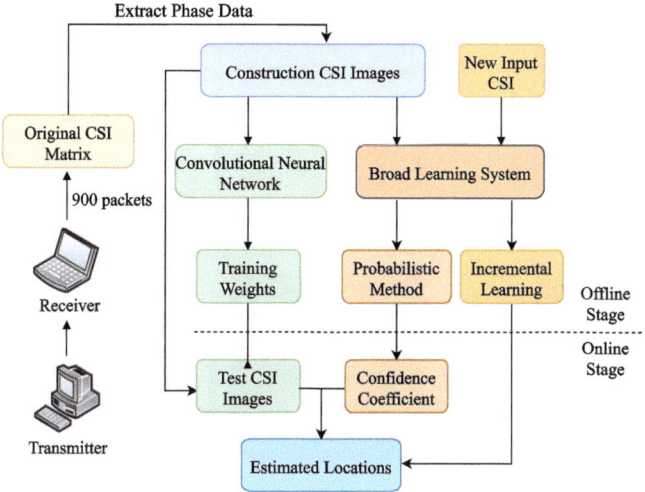

Fig. 5.5 The incremental BLS based on CSI

mechanism in the online stage to update the BLS module directly using new samples, enabling real-time adaptability with minimal overhead. Therefore, the proposed ILCL provides a lightweight and flexible method for evolving indoor environments. Experimental results verify that ILCL has superior accuracy and responsiveness compared to several existing baseline algorithms.

5.5.2 The Overview

The overall architecture of ILCL is illustrated in Fig. 5.5. To collect CSI data, we use a controlled packet transmission strategy, enabling consistent data rates and reducing offline acquisition time. At each reference point, a minimum of 3000 packets is recorded, from which 900 are selected for phase information extraction and transformed into CSI image representations. Compared to amplitude data, CSI phase measurements exhibit stronger resilience to signal blockage and capture richer spatial features.

To construct the CSI images, we take advantage of the fact that the Intel 5300 NIC outputs phase data from 30 subcarriers across three antennas. By reshaping the phase data from 900 packets, we generate 30 images per location, each with dimensions $30 \times 30 \times 3$. These images use the three antenna streams as RGB channels, forming color-like representations suitable for CNN-based training. Figure 5.6 presents examples of these CSI images from two distinct positions spaced 50 cm apart. Each heat map visualizes phase values, with rows denoting packets and columns representing subcarrier indices. Clear differences in image patterns

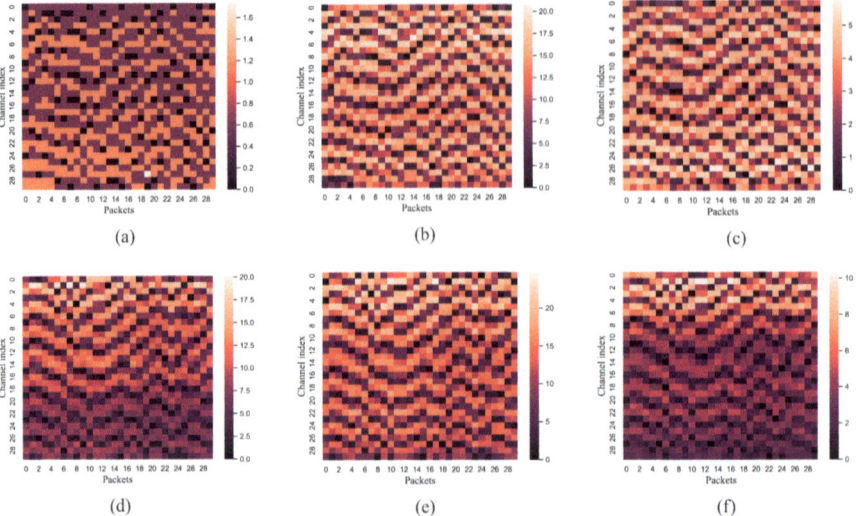

Fig. 5.6 CSI images of different RGB channels at two indoor locations. (**a**) Channel R at location 1. (**b**) Channel G at location 1. (**c**) Channel B at location 1. (**d**) Channel R at location 2. (**e**) Channel G at location 2. (**f**) Channel B at location 2

between the two locations indicate that these images capture location-specific features, making them effective as unique fingerprints for localization.

ILCL operates in two stages. In the offline phase, CSI images are used to train a CNN that extracts discriminative spatial features. These features are then passed to a probabilistic localization module built on a BLS framework. The BLS outputs classification probabilities for candidate locations, with the highest score representing the model's confidence in its prediction. A key advantage of this structure is that the BLS component supports incremental updates, allowing the model to incorporate new CSI samples efficiently without full retraining.

In the online phase, the CNN is employed to infer feature representations from incoming CSI data, and a K-means clustering algorithm is applied to refine these predictions. The estimated location is determined by integrating the clustering center with a regression estimate derived from the weighted probabilities provided by the BLS classifier. This hybrid inference mechanism balances robustness and adaptability, producing accurate localization outputs that reflect both historical training data and real-time signal characteristics.

5.5.3 Training CSI Images Based on CNN

The ILCL algorithm uses a CNN to extract informative features from the constructed CSI images, as shown in Fig. 5.7. The network is composed of multiple convo-

5.5 ILCL Algorithm

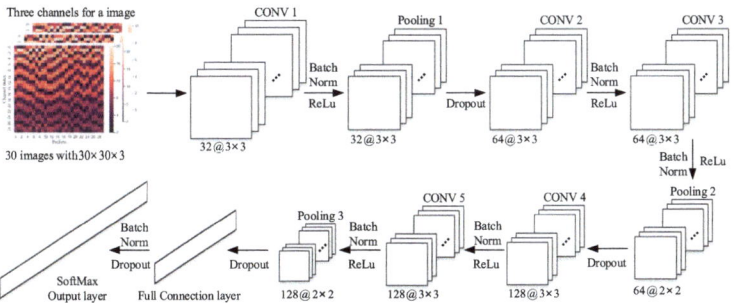

Fig. 5.7 CNN framework for CSI images

lutional layers interleaved with pooling operations, followed by a fully connected layer for final mapping. While CNN is traditionally used for classification, here they are adapted for regression-based localization. The structure is designed to learn meaningful representations from the phase-based input images.

Each convolutional layer performs localized filtering by sliding a set of kernels across the input feature maps. It captures local patterns and generates a new set of activation maps. Mathematically, the output of a convolutional unit at a given position is computed as

$$q_{j,m} = \sigma \left(\sum_{i=1}^{I} \sum_{n=1}^{F} o_{i,n+m-1} \cdot w_{i,j,n} + w_{o,j} \right), \quad (5.22)$$

where $o_{i,m}$ represents the input from the i-th feature map, $w_{i,j,n}$ denotes the convolutional kernel weights, F is the kernel size, $w_{o,j}$ is the bias term, and $\sigma(\cdot)$ is the activation function. The ReLU function is adopted to introduce non-linearity, improve sparsity, and reduce the risk of vanishing gradients. In addition, Batch normalization is further applied to accelerate convergence (Awais et al. 2021).

Pooling layers follow selected convolution stages to reduce the spatial resolution of feature maps while retaining key patterns. ILCL adopts max pooling, which selects the maximum value within a local region as

$$p_{i,m} = \max_{n=1}^{G} q_{i,(m-1) \cdot s + n}, \quad (5.23)$$

where G is the pooling window size and s is the stride. Pooling contributes to translation invariance and computational efficiency. Given the limited number of CSI images and their high similarity across locations, the risk of overfitting remains significant. To counter this, the dropout technique is integrated. During training, a portion of neuron activations (set at 25% in ILCL) are randomly disabled, encouraging the model to develop redundant and more robust representations. These inactive neurons are ignored in both forward and backward propagation, resulting in a network that learns diverse patterns and resists reliance on specific activations.

At the final stage, the fully connected layer transforms the learned abstract features into outputs suitable for localization. This stage performs a linear mapping, with parameters optimized using SGD. The network uses the cross entropy loss function to guide learning, defined as

$$E = -\sum_{j=1}^{T} y_j \log p_j, \tag{5.24}$$

where y is a one-hot encoded ground truth vector and p_j is the predicted probability, calculated via the softmax function. As shown in Fig. 5.7, each CSI image with size $30 \times 30 \times 3$ undergoes five stages of convolution and three stages of pooling. The intermediate feature maps are normalized and partially dropped out before being passed to the fully connected layer. It allows ILCL to derive robust and location-specific representations from CSI phase images, which are ultimately used to perform accurate localization through regression.

5.5.4 Estimation Location Based on A Probabilistic Method

To improve the reliability and robustness of location inference, we combine the classification results of the trained CNN model with a probabilistic estimation mechanism based on BLS. Specifically, BLS is leveraged to model the spatial relationships between CSI samples and their corresponding locations, enabling the generation of regression-based probabilistic expectations that are further fused with CNN predictions for final estimation.

After obtaining the classification probability distribution of each test sample, K-means clustering is applied to the CNN-predicted labels. Since test samples from the same location often form near-normal distributions in the embedding space, cluster centers can approximate coarse location anchors. To refine these anchors, BLS is introduced with enhanced regression capabilities. We improve the BLS model by adding an additional $(n+1)$th feature mapping node as

$$M_{n+1} = \psi(HW_{e_{n+1}} + \beta_{e_{n+1}}), \tag{5.25}$$

where H is the input matrix, $W_{e_{n+1}}$, $\beta_{e_{n+1}}$ are weights and biases, and $\psi(\cdot)$ is the activation function. This mapping introduces new variations to enrich the feature space. The corresponding enhancement nodes are constructed as

$$E_{ex_m} = [\zeta(M_{n+1}W_{ex_1} + \beta_{ex_1}), \ldots, \zeta(M_{n+1}W_{ex_m} + \beta_{ex_m})], \tag{5.26}$$

5.5 ILCL Algorithm

where $\zeta(\cdot)$ is a nonlinear activation function and m is the number of enhancement nodes.

The updated feature matrix A_{n+1}^m, which now includes the original mappings, the newly added mapping, and the corresponding enhancement nodes, is constructed as $A_{n+1}^m = [A_n^m | M_{n+1} | E_{ex_m}]$. To efficiently compute the updated model parameters without retraining from scratch, we adopt an incremental pseudoinverse-based update strategy. The pseudoinverse of A_{n+1}^m is given by

$$(A_{n+1}^m)^+ = \begin{bmatrix} (A_n^m)^+ - DB^T \\ B^T \end{bmatrix}, \quad (5.27)$$

where the intermediate matrices are defined as $D = (A_n^m)^+ [M_{n+1} | E_{ex_m}]$,

$$B^T = \begin{cases} (C)^+, & \text{if } C \neq 0 \\ (1 + D^T D)^{-1} D^T (A_n^m)^+, & \text{if } C = 0 \end{cases} \quad (5.28)$$

and $C = [M_{n+1} | E_{ex_m}] - A_n^m D$. These updates ensure numerical stability while incorporating new features, maintaining the efficiency of BLS even under continuous learning scenarios. Then, the final output weight matrix is also incrementally updated by

$$(W_{n+1}^m) = \begin{bmatrix} W_n^m - DB^T Y \\ B^T Y \end{bmatrix}. \quad (5.29)$$

After the BLS model is trained, it produces a classification probability matrix Pro_{ij}, where i indexes the test samples and j indexes the predicted label categories. For each test sample, we select the top five class probabilities and their associated location labels. A preliminary regression expectation, denoted as bls_ave_i, is calculated as a weighted average of these locations based on their class probabilities. To incorporate classification confidence into the final prediction, we extract max_pro_i, the highest probability score for each sample, and combine it with the corresponding CNN-predicted location cnn_pre_i. The final estimated location P_i is calculated as

$$P_i = bls\_ave_i \times (1 - max\_pro_i) + cnn\_pre_i \times max\_pro_i. \quad (5.30)$$

The fusion strategy adaptively balances between the regression-based estimation from BLS and the classification-based estimation from CNN, guided by the confidence of each prediction. High-confidence samples rely more on CNN outputs, while uncertain predictions are compensated by BLS regression, leading to improved robustness and accuracy in final location inference.

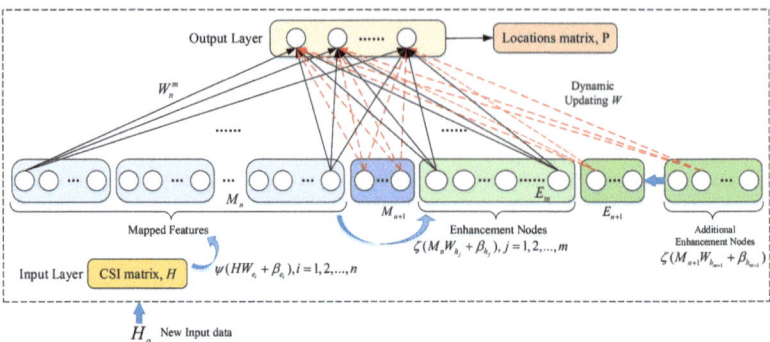

Fig. 5.8 The incremental BLS for CSI-based localization

5.5.5 Incremental Learning with New Input Data

Indoor localization systems must frequently update their fingerprint databases to adapt to dynamic environments. Traditionally, updating a CNN-based localization model with new fingerprint data requires complete retraining, which is computationally expensive (Zhu et al. 2020). To address this issue, BLS provides an efficient incremental learning mechanism that allows for fast adaptation without retraining the entire model from scratch.

As shown in Fig. 5.8, we denote the incoming new data (either a single sample or multiple samples) as H_a. The structure A_n^m represents n feature mapping nodes and m enhancement nodes in the current BLS model. When H_a arrives, new feature mappings are computed as $M_x^n = [\psi(H_a W_{e_1} + \beta_{e_1}), \ldots, \psi(H_a W_{e_n} + \beta_{e_n})]$. These mappings are concatenated with their corresponding enhancement nodes as

$$A_x = [M_x^n | \zeta(M_x^n W_{h_1} + \beta_{h_1}), \ldots, \zeta(M_x^n W_{h_m} + \beta_{h_m})]. \tag{5.31}$$

Then, the extended feature matrix for incremental learning becomes

$$^x A_n^m = \begin{bmatrix} A_n^m \\ A_x^T \end{bmatrix}. \tag{5.32}$$

Following the update rule in Eqs. (5.27), (5.28), and (5.29), the pseudoinverse of the extended matrix is updated as $(^x A_n^m)^+ = [(A_n^m)^+ - BD^T | B]$, where $D^T = A_x^T (A_n^m)^+$, and the matrix B^T is given by

$$B^T = \begin{cases} (C)^+ & if \quad C \neq 0 \\ (1 + D^T D)^{-1} (A_n^m)^+ D & if \quad C = 0 \end{cases} \tag{5.33}$$

5.5 ILCL Algorithm

Algorithm 4 Incremental learning process of ILCL

1: **Input:** Original training data, test data, new input samples H_a, BLS parameters
2: **Output:** Estimated location for the new input samples
3: Randomly initialize weights and biases
4: Compute initial feature mappings M_n and enhancement nodes E_m
5: Initialize number of feature mappings n and enhancement nodes m
6: **while** Stopping criterion is not met **do**
7:     **if** Adding enhancement nodes **then**
8:         Randomly initialize weights and biases for new enhancement nodes
9:         Compute $(A_n^{m+1})^+$ and W_n^{m+1}
10:         $m \leftarrow m + 1$
11:     **else if** Adding additional feature mappings **then**
12:         Randomly initialize weights and biases for new feature mappings
13:         Compute $(A_{n+1}^m)^+$ and W_{n+1}^m
14:         $n \leftarrow n + 1$
15:     **else**
16:         **Incorporate new input samples H_a**
17:         Compute updated mapping A_x
18:         Augment the mapping matrix: $^xA_n^m$
19:         Compute the pseudoinverse $(^xA_n^m)^+$
20:         Update the weights: $^xW_n^m$
21:     **end if**
22: **end while**
23: Use the trained incremental model to predict the test data
24: Estimate the location of new samples using cluster-based decision
25: **Return** Estimated location

---

with $C = A_x^T - D^T A_n^m$. Finally, the output weight matrix is updated as $^xW_n^m = W_n^m + (Y_a^T - A_x^T W_n^m)B$, where Y_a denotes the label vector of the newly added input data. Since these new samples are not part of the original CNN training set, their final location estimates are obtained by clustering the predictions of the updated BLS model. And the above processing is listed in Algorithm 4.

5.5.6 Performance Evaluation

The performance of the ILCL algorithm is still evaluated in the two scenarios. And the comparisons methods including CiFi (Wang et al. 2020), AF-DCGAN (Li et al. 2021), EnsemLoca (Wu et al. 2021), BLS-Location (Zhu et al. 2021), and SWIM (Chen et al. 2019a).

(1) Localization Comparisons The results of the experiment are shown in Table 5.4, and Fig. 5.9. In both test areas, ILCL consistently achieves the highest localization accuracy and the lowest testing time. In Area one, it record a mean error of 2.38 m with a testing time of 0.002 s, while in Area two, it reaches 1.29 m with only 0.001 s required for inference. Compared to other methods, such as EnsemLoca with 2.56 and 1.35 m mean errors in the two areas respectively, ILCL

Table 5.4 Localization accuracy and testing time of ILCL

Algorithms	Area one			Area two		
	Mean errors (m)	Std (m)	Testing time (s)	Mean errors (m)	Std (m)	Testing time (s)
ILCL	**2.38**	1.75	**0.002**	**1.29**	1.03	**0.001**
CiFi (Wang et al. 2020)	4.57	2.21	2.001	3.60	2.02	1.190
AF-DCGAN (Li et al. 2021)	5.74	3.63	51.269	4.08	2.80	28.811
EnsemLoca (Wu et al. 2021)	2.56	**1.43**	14.925	1.35	**0.89**	8.315
BLS-Location (Zhu et al. 2021)	4.43	1.97	0.006	2.55	1.19	0.004
SWIM (Chen et al. 2019a)	5.42	2.88	1.609	3.35	1.86	0.511

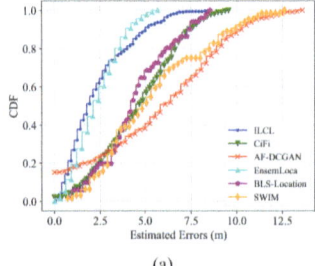
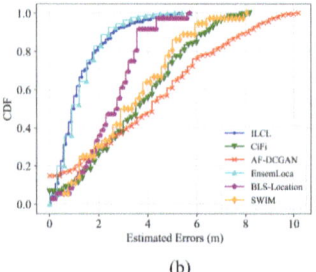

Fig. 5.9 The localization performance comparisons. (**a**) Area one. (**b**) Area two

shows both better precision and greater efficiency. Although EnsemLoca maintains the lowest standard deviation in both settings, its testing time remains considerably higher. Methods like AF-DCGAN and SWIM show poor accuracy and longer inference durations, making them less suitable for real-time applications.

(2) Ablation Experiments To investigate the influence of antenna count on localization accuracy, we utilize the CSI data captured by the Intel 5300 NIC, which records signals from three antennas. These multi-antenna signals are treated as RGB-like channels when constructing CSI images. Specifically, we generate 30×30 phase images with a third dimension of 1, 2, or 3 to represent the use of one, two, or all three antennas. Each sampling location provides 900 packets, and separate models are trained for each antenna setting using the same hyperparameters and a 4:1 cross-validation strategy to ensure a fair comparison. As illustrated in Fig. 5.10a, utilizing all three antennas yields the highest localization accuracy in both scenarios, as the additional spatial information enhances feature extraction. Nevertheless, ILCL still performs well even with a single antenna, achieving mean errors up to 2.94 and 1.39 m in the two environments. It verifies the model's robustness and

5.5 ILCL Algorithm

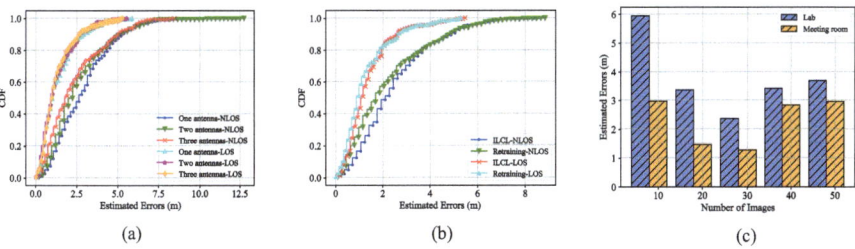

Fig. 5.10 The performance comparisons of different settings. (**a**) Antenna. (**b**) Incremental learning. (**c**) Image number

its ability to extract meaningful features from stable phase information despite the reduced input dimensionality.

Next, the ILCL includes an incremental learning strategy based on BLS, which allows the model to be updated with new data while preserving the previously trained parameters. In contrast, CNN-based localization methods require complete retraining once new data are introduced, which is both time-consuming and computationally expensive. To evaluate the effectiveness of the incremental learning mechanism, we simulate updates by designating the last two rows of sampling locations as newly collected data. This results in 46 new locations out of 317 in Area one, and 22 out of 176 in Area two. Figure 5.10b presents a comparison between ILCL and retrained CNN models, focusing on localization accuracy and training efficiency. Although ILCL shows slightly higher mean errors, reaching 2.54 m in the lab and 1.40 m in Area two, it achieves substantial reductions in training time while maintaining reliable performance. It show ILCL's suitability for real-world applications that require rapid adaptation to new data.

Finally, to evaluate how the number of CSI images influences localization accuracy, we construct five datasets for each environment, varying the number of CSI images per location from 10 to 50, with each image sized at $30 \times 30 \times 3$. The transmitter operates at a consistent rate during data collection, and consecutive packets are used to generate the CSI images to ensure fairness. As shown in Fig. 5.10c, using only 10 images per location leads to underfitting, resulting in the highest mean errors of 5.95 m in Area one and 2.98 m in Area two. When the number of CSI images increases to 50, the mean errors also rise, which is likely caused by overfitting. The best localization performance is achieved with 30 CSI images, where the lowest mean errors of 2.38 m and 1.29 m are obtained in Area one and Area two, respectively.

To summarize, this chapter introduces ILCL, a localization algorithm that leverages CSI phase images and incremental learning based on BLS. By transforming the more stable phase information into images, ILCL enables effective feature extraction through a multi-layer CNN. A probabilistic approach is incorporated within the BLS framework to estimate location confidence. Unlike conventional methods, ILCL can incorporate new data without retraining the entire model, significantly reducing offline training time. Experimental results in two typical indoor environments demonstrate that ILCL achieves higher accuracy than several baseline methods.

References

Awais M, Iqbal MTB, Bae SH (2021) Revisiting internal covariate shift for batch normalization. IEEE Trans Neural Netw Learn Syst 32(11):5082–5092
Bahl P, Padmanabhan VN, Bahl V, Padmanabhan V (2000) RADAR: an in-building RF-based user location and tracking system. In: IEEE international conference on computer communications, pp 775–784
Chen CP, Liu Z (2017) Broad learning system: an effective and efficient incremental learning system without the need for deep architecture. IEEE Trans Neural Netw Learn Syst 29(1):10–24
Chen M, Liu K, Ma J, Gu Y, Dong Z, Liu C (2019a) SWIM: speed-aware WiFi-based passive indoor localization for mobile ship environment. IEEE Trans Mobile Comput 20:765–779
Chen N, Qiu T, Zhou X, Li K, Atiquzzaman M (2019b) An intelligent robust networking mechanism for the internet of things. IEEE Commun Mag 57(11):91–95
Daniel D (2012) An implementation of the Kalman Filter, Kalman Smoother, and EM algorithm in Python. https://github.com/pykalman/pykalman/
Djuric PM, Kotecha JH, Zhang J, Huang Y, Ghirmai T, Bugallo MF, Miguez J (2003) Particle filtering. IEEE Signal Process Mag 20(5):19–38
Fernández DN (2019) Implementation of a WiFi-based indoor location system on a mobile device for a university area. In: IEEE XXVI international conference on electronics, electrical engineering and computing, pp 1–4
Khairuddin AR, Talib MS, Haron H (2015) Review on simultaneous localization and mapping (SLAM). In: 2015 IEEE international conference on control system, computing and engineering (ICCSCE), IEEE, New York, pp 85–90
Li Q, Qu H, Liu Z, Zhou N, Sun W, Sigg S, Li J (2021) AF-DCGAN: amplitude feature deep convolutional GAN for fingerprint construction in indoor localization systems. IEEE Trans Emerg Top Comput Intell 5(3):468–480
Lu Y, Ma H, Smart E, Yu H (2021) Real-time performance-focused localization techniques for autonomous vehicle: a review. IEEE Trans Intell Transp Syst 23(7):6082–6100
Malkov YA, Yashunin DA (2018) Efficient and robust approximate nearest neighbor search using hierarchical navigable small world graphs. IEEE Trans Pattern Anal Mach Intell 42(4):824–836
Niu S, Liu Y, Wang J, Song H (2021) A decade survey of transfer learning (2010–2020). IEEE Trans Artif Intell 1(2):151–166
Ren P, Xiao Y, Chang X, Huang PY, Li Z, Gupta BB, Chen X, Wang X (2021) A survey of deep active learning. ACM Comput Surv 54(9):1–40
Roy P, Chowdhury C (2021) A survey of machine learning techniques for indoor localization and navigation systems. J Intell Robot Syst 101(3):63
Sagi O, Rokach L (2018) Ensemble learning: a survey. Wiley Interdisciplinary Rev Data Min Knowl Discov 8(4):e1249
Simon D (2001) Kalman filtering. Embed Syst Program 14(6):72–79
Taniuchi D, Maekawa T (2014) Robust Wi-Fi based indoor positioning with ensemble learning. In: 2014 IEEE 10th international conference on wireless and mobile computing, networking and communications (WiMob), IEEE, New York, pp 592–597
Umer A, Müürsepp I, Alam MM, Wymeersch H (2025) Reconfigurable intelligent surfaces in 6G radio localization: a survey of recent developments, opportunities, and challenges. IEEE Commun Surv Tuts. https://doi.org/10.1109/COMST.2025.3536517

References

Van de Ven GM, Tuytelaars T, Tolias AS (2022) Three types of incremental learning. Nat Mach Intell 4(12):1185–1197

Wang X, Wang X, Mao S (2020) Deep convolutional neural networks for indoor localization with CSI images. IEEE Trans Netw Sci Eng 7(1):316–327

Wu C, Qiu T, Zhang C, Qu W, Wu DO (2021) Ensemble strategy utilizing a broad learning system for indoor fingerprint localization. IEEE Internet Things J 9(4):3011–3022

Youssef M, Agrawala A (2005) The Horus WLAN location determination system. In: Proceedings of the 3rd international conference on mobile systems, applications, and services, pp 205–218

Zheng H, Gao M, Chen Z, Liu XY, Feng X (2019) An adaptive sampling scheme via approximate volume sampling for fingerprint-based indoor localization. IEEE Internet Things J 6(2):2338–2353

Zhang B, Sifaou H, Li GY (2023) CSI-fingerprinting indoor localization via attention-augmented residual convolutional neural network. IEEE Trans Wireless Commun 22(8):5583–5597

Zhou K, Hou Q, Wang R, Guo B (2008) Real-time KD-tree construction on graphics hardware. ACM Trans Graph 27(5):1–11

Zhu X, Qu W, Qiu T, Zhao L, Atiquzzaman M, Wu DO (2020) Indoor intelligent fingerprint-based localization: principles, approaches and challenges. IEEE Commun Surv Tuts 22(4):2634–2657

Zhu X, Qiu T, Qu W, Zhou X, Atiquzzaman M, Wu D (2021) BLS-location: a wireless fingerprint localization algorithm based on broad learning. IEEE Trans Mobile Comput 22(1):115–128

Zhu X, Qu W, Zhou X, Zhao L, Ning Z, Qiu T (2022) Intelligent fingerprint-based localization scheme using CSI images for internet of things. IEEE Trans Netw Sci Eng 9(4):2378–2391

Open Access This chapter is licensed under the terms of the Creative Commons Attribution 4.0 International License (http://creativecommons.org/licenses/by/4.0/), which permits use, sharing, adaptation, distribution and reproduction in any medium or format, as long as you give appropriate credit to the original author(s) and the source, provide a link to the Creative Commons license and indicate if changes were made.

The images or other third party material in this chapter are included in the chapter's Creative Commons license, unless indicated otherwise in a credit line to the material. If material is not included in the chapter's Creative Commons license and your intended use is not permitted by statutory regulation or exceeds the permitted use, you will need to obtain permission directly from the copyright holder.

Chapter 6
Conclusion

Abstract This chapter provides a comprehensive summary of the key insights and innovations presented throughout the book, with a focus on the evolution of CSI-based indoor localization techniques within ISAC systems. It revisits the core challenges of indoor localization, such as inefficient data collection, lack of intelligent updates, and sensitivity to environmental dynamics, and highlights how machine learning-based methods can effectively address these limitations. The chapter synthesizes the strategies introduced in earlier chapters, including automated CSI collection using A3C-IPP and CPPU, intelligent fingerprint database updates through generative CSI and DBLG, and real-time localization algorithms such as BLS-Location and ILCL.

In addition, the chapter outlines the broader vision for future research in 6G, where localization is expected to become a fundamental capability of intelligent communication systems. Key directions include the integration of localization, sensing, and computing, the adoption of AI for system self-adaptation, and the application of emerging technologies such as RIS and terahertz communication.

Keywords 6G · Integrated sensing and communication · Feature research

6.1 Summary of Key Insights

Indoor localization, as a fundamental enabler for IoT applications and the advancement of ISAC systems, is becoming increasingly important. With the continuous evolution of ISAC technologies, there is a growing demand for more efficient data acquisition, intelligent data updating, and more accurate localization estimation. These capabilities are vital for enhancing system sensing performance, optimizing resource allocation, and improving overall service quality. While Global Navigation Satellite Systems, such as GPS, are highly effective for outdoor localization, their performance in indoor environments is often compromised by signal blockages and multipath effects, limiting their applicability. Therefore, the development of robust indoor localization technologies is essential for fully realizing the potential of 6G ISAC networks.

However, current indoor localization methods still face a range of significant challenges. Traditional CSI data collection largely depends on manual operations, which are not only time-consuming and labor-intensive but also inadequate for supporting large-scale deployment needs. At the same time, existing systems often lack intelligent mechanisms for CSI data updates, making it difficult to promptly capture and reflect environmental changes, which in turn leads to rapid declines in localization performance. Furthermore, the prevalence of signal interference and multipath effects indoors continues to constrain localization accuracy. Addressing these pressing issues calls for the development of more efficient, intelligent, and precise indoor localization solutions.

Through theoretical analysis and extensive practical validation, this book offers a comprehensive exposition of the latest advancements and research developments in machine learning-based CSI localization systems. It places particular emphasis on the adaptability and effectiveness of these methods in complex indoor environments and proposes systematic strategies for improvement across multiple dimensions. The first two chapters highlight the inherent challenges of indoor localization and the pivotal role of CSI in wireless communications. They identify three main research directions: efficient CSI collection, intelligent CSI updating, and accurate localization, which collectively establish a strong foundation for the chapters that follow. Furthermore, the book systematically examines machine learning-based indoor localization techniques and corresponding optimization strategies, reinforcing both the theoretical framework and practical methodologies for intelligent indoor localization.

During the data collection phase, the focus is placed on offline data acquisition, with a comparative analysis of manual versus automated methods, as well as device-based versus device-free approaches. To improve the efficiency and quality of data collection, the book introduces two optimization algorithms: A3C-IPP and CPPU. For intelligent offline data updating, the book systematically presents techniques for CSI prediction, completion, and enhancement, alongside strategies for constructing and refining fingerprint databases. It further proposes the generative CSI algorithm and the DBLG algorithm to enhance system accuracy and robustness. In addressing the challenges of dynamic environments, the book conducts an analysis and proposes the BLS-Location and ILCL algorithms. These methods effectively balance speed and accuracy, providing practical and scalable solutions for intelligent indoor localization systems.

We hope this book will serve as a valuable reference for researchers, engineers, and practitioners engaged in the study and application of intelligent localization technologies, helping them stay at the forefront of this rapidly evolving field and address emerging challenges. At the same time, it is intended to support students and newcomers who are interested in entering this area, offering a clear and structured foundation to build their knowledge and inspiring further exploration and innovation. We believe that this work will not only provide immediate guidance for current research and practice but also lay a strong foundation for future developments.

6.2 Future Directions: 6G and Beyond

For the 6G ISAC, localization will shift from being a secondary function of communication systems to a core capability in intelligent environments, supporting the deployment of immersive experiences, real time sensing, and context aware services. Meeting the demands for centimeter level accuracy, strong adaptability in dynamic conditions, and reliable services with minimal delay will be essential for enabling the next generation of applications.

Future research should continue to drive the deep convergence of localization, communication, sensing, and computing, moving toward truly unified ISAC systems. Machine learning and artificial intelligence will play an increasingly pivotal role, enabling systems to autonomously adapt to environmental changes, anticipate dynamic conditions, and optimize overall performance without human intervention. As advances in algorithms and hardware technologies accelerate, these systems will be better equipped to operate efficiently in complex, dynamic, and resource-constrained environments. At the same time, emerging technologies such as Reconfigurable Intelligent Surfaces (RIS), terahertz communications, and quantum sensing are poised to fundamentally reshape existing indoor localization paradigms, opening new pathways for technological innovation and broader application development.

Building on the theoretical framework and practical methodologies outlined in this book, we envision future intelligent localization systems achieving self-evolution, full scalability, and seamless integration with next-generation network infrastructures. Such advancements will greatly enhance the overall performance, reliability, and adaptability of localization systems. Achieving this vision will demand continuous progress through deep interdisciplinary collaboration across wireless communications, machine learning, signal processing, and systems engineering. Through these collective efforts, the immense potential of 6G and future technologies can be fully realized, driving the rapid growth of next-generation intelligent environments and diverse application ecosystems.

Open Access This chapter is licensed under the terms of the Creative Commons Attribution 4.0 International License (http://creativecommons.org/licenses/by/4.0/), which permits use, sharing, adaptation, distribution and reproduction in any medium or format, as long as you give appropriate credit to the original author(s) and the source, provide a link to the Creative Commons license and indicate if changes were made.

The images or other third party material in this chapter are included in the chapter's Creative Commons license, unless indicated otherwise in a credit line to the material. If material is not included in the chapter's Creative Commons license and your intended use is not permitted by statutory regulation or exceeds the permitted use, you will need to obtain permission directly from the copyright holder.

MIX
Papier aus verantwortungsvollen Quellen
Paper from responsible sources
FSC® C105338

If you have any concerns about our products,
you can contact us on
ProductSafety@springernature.com

In case Publisher is established outside the EU,
the EU authorized representative is:
**Springer Nature Customer Service Center GmbH
Europaplatz 3, 69115 Heidelberg, Germany**

Printed by Libri Plureos GmbH
in Hamburg, Germany